T0331084

Hands-On Data Science for Librarians

Librarians understand the need to store, use and analyze data related to their collections, patrons and institution, and there has been consistent interest over the last 10 years to improve data management, analysis, and visualization skills within the profession. However, librarians find it difficult to move from out-of-the-box proprietary software applications to the skills necessary to perform the range of data science actions in code. This book will focus on teaching R through relevant examples and skills that librarians need in their day-to-day lives that includes visualizations but goes much further to include web scraping, working with maps, creating interactive reports, machine learning, and others. While there's a place for theory, ethics, and statistical methods, librarians need a tool to help them acquire enough facility with R to utilize data science skills in their daily work, no matter what type of library they work at (academic, public or special). By walking through each skill and its application to library work before walking the reader through each line of code, this book will support librarians who want to apply data science in their daily work. *Hands-On Data Science for Librarians* is intended for librarians (and other information professionals) in any library type (public, academic or special) as well as graduate students in library and information science (LIS).

Key Features:

- Only data science book available geared toward librarians that includes step-by-step code examples
- Examples applicable for all library types (public, academic, special)
- Relevant datasets
- Accessible to non-technical professionals
- Focused on job skills and their applications

Hands-On Data Science for Librarians

Sarah Lin
Dorris Scott

CRC Press
Taylor & Francis Group
Boca Raton London New York

CRC Press is an imprint of the
Taylor & Francis Group, an **informa** business
A CHAPMAN & HALL BOOK

Designed cover image: https://www.shutterstock.com/image-photo/round-bookshelf-public-library-170760092

First edition published 2023
by CRC Press
6000 Broken Sound Parkway NW, Suite 300, Boca Raton, FL 33487-2742

and by CRC Press
4 Park Square, Milton Park, Abingdon, Oxon, OX14 4RN

CRC Press is an imprint of Taylor & Francis Group, LLC

ISBN: 978-1-032-10999-2 (hbk)
ISBN: 978-1-032-08082-6 (pbk)
ISBN: 978-1-003-21801-2 (ebk)

DOI: 10.1201/9781003218012

Typeset in Latin Modern font
by KnowledgeWorks Global Ltd.

Publisher's note: This book has been prepared from camera-ready copy provided by the authors.

For Will, Bennett, and Bridget—S.L.

For Mom–D.S.

Contents

List of Figures

List of Tables

Preface

Resources to learn R are all over the internet and most libraries. However, easy access to resources doesn't mean it's easy to learn to do data science in R. This book spends time on an introduction to R and basic data cleaning tasks that are taught elsewhere because we want to provide a gentle, low-stress introduction to key aspects of data science using R. Librarians have varied backgrounds, but for most of us, rigorous education in mathematics, statistics, and computer science are not part of our expertise. That doesn't mean we can't learn to code or do data science in code. Based on our own experiences, we are particularly concerned that you, our reader, are able to access the content in this book with minimal frustration, exasperation, and despair.

Using resources at the end of each chapter, in the appendix, and in the bibliography of this book will provide you with next steps to further your data science skills beyond this introductory text. With a basic foundation in data science skills, any of the resources we link to should be comprehensible, if challenging. We wish you well on your data science journey!

What you'll need

Access to a personal computer (desktop or laptop) with permission to install programs, such as the Chrome web browser and extensions.

Software information, notes, and conventions

We used the *knitr* package and the *bookdown* package to compile this book. Our R session information is shown below:

```
xfun::session_info()
```

```
## R version 4.2.1 (2022-06-23)
## Platform: x86_64-apple-darwin17.0 (64-bit)
## Running under: macOS Big Sur ... 10.16
##
## Locale: en_US.UTF-8 / en_US.UTF-8 / en_US.UTF-8 / C / en_US.UTF-
8 / en_US.UTF-8
##
## Package version:
##    base64enc_0.1.3 bookdown_0.29   bslib_0.4.0
##    cachem_1.0.6    cli_3.4.1       compiler_4.2.1
##    digest_0.6.29   evaluate_0.16   fastmap_1.1.0
##    fs_1.5.2        glue_1.6.2      graphics_4.2.1
##    grDevices_4.2.1 highr_0.9       htmltools_0.5.3
##    jquerylib_0.1.4 jsonlite_1.8.0  knitr_1.40
##    magrittr_2.0.3  memoise_2.0.1   methods_4.2.1
##    R6_2.5.1        rappdirs_0.3.3  rlang_1.0.6
##    rmarkdown_2.17  rstudioapi_0.14 sass_0.4.2
##    stats_4.2.1     stringi_1.7.8   stringr_1.4.1
##    tinytex_0.42    tools_4.2.1     utils_4.2.1
##    xfun_0.33       yaml_2.3.5
```

Package names are in italic text (e.g., *rmarkdown*), and inline code and filenames are formatted in a typewriter font (e.g., `knitr::knit('foo.Rmd')`). Function names are followed by parentheses (e.g., `bookdown::render_book()`).

We use the assignment (<-) operator in all code chunks to assign and store objects in this book, but you can also use the equals sign (=).

In 2022, the company RStudio, PBC changed its name to Posit, PBC. The open source IDE created by this company is now known as either "the RStudio IDE", or simply "RStudio." We use both terms interchangeably in this book.

Acknowledgments

We would like to thank the following individuals for the gift of their time, insight, and support while we wrote and edited this book: Greg Wilson, Rafael Leonardo Da Silva, Emily Nimsakont, Myfanwy Johnston, Patrick Alston, Luke Johnston, Carl Howe, Emil Hvitfeldt, and Julia Silge.

About the Authors

Sarah Lin is the Senior Information and Content Architect at MongoDB and previously managed the Enterprise Information Management team at Posit, PBC. A graduate of the University of Illinois iSchool, Sarah worked as a technical services librarian in many different library types (academic, special, medical, & law) before moving into corporate librarianship and information management. Her professional interests include findability and metadata, and how metadata enables findability. Sarah is a certified Carpentries instructor and received her undergraduate degree in African/African-American Studies & Anthropology from the University of Chicago. She didn't know anything about coding in any programming language before joining Posit. Her website, created with R code, is at http://sarah.rbind.io.

Dorris Scott is currently the Academic Director of Data Studies at Washington University in St. Louis' school of continuing education and professional studies, known as University College. As Academic Director of Data Studies, Dorris develops new curriculum, certificate and degree programs related to data analytics, data science, and geographic information systems (GIS) and also builds local and regional partnerships in alignment with University College's strategic vision. Previously, Dorris was Geographic Information Systems (GIS) Librarian and Social Science Data Curator at University Libraries at Washington University and provided consultation on projects that use geospatial data along with providing training in various GIS software, programming applications of geospatial data, and data management. As part of this position, Dorris also served as a liaison between Washington University Libraries and social science departments and assisted faculty with their data needs such as data management and data curation. Dorris received a PhD in Geography from the University of Georgia, with a specialization in GIS applications for public health.

1

Introduction

1.1 What Is Data Science?

Data science degree and certificate programs have sprouted at academic institutions around the US, while books, articles, and conference programs about data and how to analyze it regularly appear in library conference programs and educational events. The increased visibility of data science belies the fact that data science has been around for a while. Indeed, data collection and the need to make sense of it are not new. R, the programming language used in this book, has been around for decades. However, experts have some back-and-forth about the discipline of data science and its relationship to other subjects.

Rather than take sides, this book takes a broad view of what constitutes data science and highlights five interdependent elements. These include both **mathematics** and **statistics** on the computational side. With or without a graphical user interface, data science is made real through **computer programming**. Practitioners of data science bring extensive **subject matter knowledge**. Their expertise enables them to communicate their conclusions through data **visualizations**, often providing pictures that speak louder than numbers.

Data science is a discipline that extracts knowledge from data in various fields, including librarianship. While data science can help make decisions, it is not a substitute for human decision-making. It can provide insights and generalizations from collected observations (data). Aspects of some subjects remain unquantifiable yet comprehensible to human interpretation. Data analysis is fallible; it requires data science practitioners to bring their expertise to bear on interpretation and decision-making.

Whether we realize it or not, data science is a broad discipline that saturates our professional lives. For academic librarians, faculty, staff, and students learn and perform data science tasks daily, such as data cleaning, management, and visualizations. This occurs in computational science disciplines as well as the biological, physical, social sciences, and even in the humanities. In addition,

DOI: 10.1201/9781003218012-1

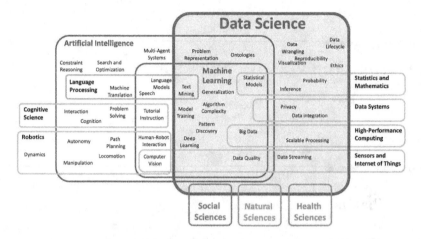

FIGURE 1.1 Data science as discipline diagram, Data Science Program, Viterbi School of Engineering, University of Southern California, http://data science.usc.edu, 2021.

librarians can act as data curators who help researchers publish or deposit their data to data repositories and academic journals.

Corporations and other institutions with special libraries likely have teams using many tools to analyze the market or user behavior. Predictive text in search engines relies upon text mining and machine learning. Humanities and social science professionals use maps, analysis, web scraping, and text mining to create and analyze datasets. These disciplines need to communicate their findings through written reports and dashboards for their stakeholders and constituents. Data science also permeates the public sphere. Users are subject to machine learning algorithms in their daily lives within loan applications, resume screenings, social media feeds, news visualizations, public health data, social services eligibility, and medical care. Public librarians interact with patrons whose complex information needs may result from how data science impacts their lives. Data literacy is required when data science provides input for human decisions, particularly when those decisions affect others' well-being.

1.2   Why Learn Data Science?

Librarians have long collected metrics about their collections and their patrons. However, the pervasiveness of data collection and the need to justify or rationalize library expenditures creates an environment that data science can exploit in the best interests of library and information professionals. Because librarians are both consumers of data and teachers of data literacy, they must acquire skills to perform data science and interrogate data analyses to determine their veracity.

Data literacy is the ability to read, interpret, and analyze data, and it is a requirement when people use data to distort the truth[1]. Unfortunately, data literacy is both a necessary and frequently needed skill. Data science enables data literacy and democratizes access to the source material; so much of our personal and professional lives are affected by data, whether created or influenced by data-driven decision-making. Data provides valuable information to help experts make decisions. Beyond just the economy, so much in our society rewards data literacy and penalizes the illiterate. Because of this, data is too valuable to be left only to data scientists, computer scientists, or statisticians. Instead, subject experts need to learn to code because they know their data best and are best suited to analyze it and draw healthy and accurate conclusions. Your professional expertise lets you ask the right questions and interpret meaning from the data. When experts in their field add data science skills to their repertoire, data science is further democratized[2], and data-driven decisions are more impactful.

1.3 Why Use Code?

Ever the proponents of literacy, librarians have embraced data literacy and data-driven decision-making for many years. Conference sessions to improve both data collection and analytics presentation abound. When data skills are adopted, it is usually in the context of a commercial spreadsheet or analytics program. Learning to code is not as common among library and information professionals; this book argues that learning to code is doable and provides increased utility and impact. In the long run, learning a programming language for data science is best because it is accessible to all, ensures data analysis is reproducible, and it is future-proof as applications change.

If we define programming as being able to talk to computers in a language they understand, then most librarians have already done that and are probably quite good at it. Technical services and cataloging librarians will be familiar

---

[1] https://royalsocietypublishing.org/doi/10.1098/rsos.190161
[2] https://posit.co/resources/videos/data-science-education-in-2022/

with MARC (Machine Readable Cataloging), the special syntax libraries use to catalog their collections so that computer software can read it. More commonly, if you've written formulas in a spreadsheet application, you've dabbled in the basics of computer programming. However, learning to code offers far greater applications and versatility than a spreadsheet application.

The core benefits of doing data science in code are interoperability and reproducibility. Many academic librarians will be familiar with FAIR Principles[3] through their data curation work; this initiative focuses on making information findable, accessible, interoperable, and reproducible. Doing data science in code ensures that data and data analysis are both interoperable and reproducible, neither of which is possible with proprietary software applications.

Interoperability requires that other librarians who may have completely different software applications on their computers would be able to run anyone else's code. The R programming language is an open-source tool that is free to anyone across the globe and provides transparent data analysis. Additionally, platform-agnostic tools like coding can bring together the output of multiple commercial products to rationalize and analyze the data together.

Reproducibility is closely related to interoperability because code should run on any application configuration. Still, the analysis must be able to be re-run by another person and get the same results. In the past few years, there have been stories in the news about errors in spreadsheet applications that allowed researchers to draw erroneous conclusions. In one case, years of austerity measures around the globe rested on one economics research paper that was missing a few values for some variables[4]. Using code allows researchers to combine their data, code, and analysis, providing transparency into the process of data science. Unfortunately, there have other examples of reproducibility problems in various scientific disciplines: physics[5], psychology[6], and medical research[7] as well. A librarian will need to re-run their analyses on new iterations of data without replicating the data cleaning and analysis steps manually. Thankfully, code can be run repeatedly with new data as input, saving hours and hours while repeating each step precisely. The ultimate benefit of doing data science using computer programming languages is the ability to share raw data and the steps for analysis.

---

[3] https://www.go-fair.org/

[4] https://www.businessinsider.com/thomas-herndon-michael-ash-and-robert-pollin-on-reinhart-and-rogoff-2013-4

[5] https://physicstoday.scitation.org/do/10.1063/PT.6.1.20180822a/full/

[6] https://www.science.org/doi/10.1126/science.aac4716

[7] https://journals.plos.org/plosbiology/article?id=10.1371/journal.pbio.1002165

1.4 Vignette

This book creates an overarching narrative that presents realistic code examples and valuable outputs centered around a hypothetical outreach librarian in St. Louis, MO. Envision that you are this outreach librarian and you want to create a partnership with community institutions to address unemployment in St. Louis. Your goal is to present a report to stakeholders at the library and within the community that analyzes several data sources related to employment and unemployment in St. Louis. You will utilize different data science skills to compile the report. Each chapter in this book will touch on a different aspect of the report, building upon each other to learn data science and code each analytical section in R.

The reader is invited to inhabit the role of this librarian, who we will address as "you" throughout the book as we introduce each chapter with a scenario that describes what the librarian is trying to accomplish with each data science skill.

1.5 Structure of This Book

In pursuit of data to justify a community partnership, you will learn R in incremental steps with a topic for each chapter that will produce one aspect of the final report. This book isn't an exhaustive textbook on R or data science but rather a guidebook through the central functional practices of data science in R. The focus is on immediately applicable skill acquisition made easier through library-specific hypothetical tasks. The chapter topics include the following:

1. Use RStudio to code in R
2. Learn to clean data using code
3. Plot basic visualizations
4. Scrape websites using code
5. Visualize data using maps
6. Use code to mine textual data
7. Publish your code using R Markdown
8. Communicate your findings via flexdashboard

9. Let stakeholders draw their conclusions from an interactive Shiny application
10. understand how artificial intelligence intersects with employment by understanding how machine learning works.

To expand on this list, the first two chapters explain R, the RStudio IDE used to program in R, and how to get started cleaning data. In any data-related project, cleaning data is the first and often the most time-consuming task. Chapters 3 through 9 teach different data science skills: plots/graphs, web scraping, geographic visualizations, text mining, publishing, dashboards, and interactive web applications. The final chapter covers machine learning, explaining the construction of algorithms and their implications for librarians who interact with them. An explanation of how resumé screening software uses machine learning to accept or reject job applications ties how machine learning works with experiences job seekers have through the prospective outreach partnership.

1.6　Who This Book Is for

The anticipated audience for this book is all librarians and information professionals interested in learning data science and applying it to their everyday jobs. Public, academic, medical, legal, special, and corporate librarians can all put the data science skills taught in this book to use in their daily work. The book has been designed with examples adaptable to many job positions and library types, creating a practical introduction to primary data science skills needed in a professional setting. This book does not include in-depth explanations of particular R packages, the statistical and mathematical principles behind package functions, or theoretical foundations of different analysis types. There are several related topics that, while not required, are helpful to learn alongside or following this book. Appendix includes those topics and resources to learn more about them.

2

Using RStudio's IDE

2.1 Learning Objectives

1. Use your computer knowledge to install RStudio.
2. Describe the function of each pane in the IDE.
3. Modify IDE settings to your liking.
4. Use the IDE to import a tabular data file.

2.2 Terms You'll Learn

- integrated development environment (IDE)
- package
- tidyverse
- session
- working directory

2.3 Scenario

You want to use R to do data science and publish a data-based report to support your outreach efforts, but you don't know how to code in R or get started.

DOI: 10.1201/9781003218012-2

2.4 Introduction

This chapter aims to get you up and running with programming in R using the RStudio **Integrated Development Environment**, or IDE, which is generally referred to as "RStudio." An IDE is a computer program that makes it easier to code; while you can use your computer's command line[1] or UNIX shell[2] interface to code, the graphical user interface of an IDE makes it a lot more accessible. The distinction between coding in the command line or using an IDE is a lot like the difference between finding stored files in the command line or using Finder/File Explorer on your work or personal computer. While there are some scenarios where using the command line makes the most sense, for the day-to-day, most computer users use the Finder/File Explorer to more easily navigate through their files and data. IDEs are very common in computer programming, and many different applications exist. We're using RStudio because it was designed specifically for R, though you can use it to program in Python and other languages. It is free and open-source, and using it to program in R is a widely used way to wrangle and interpret data. We will also cover the basics of R as a programming language, and a widely used core of packages called the Tidyverse and then install RStudio to get started with R.

2.5 What is R?

Version 1.0 of the R programming language was released publicly in 2000[3], five years after initial distribution as open-source software. The intellectual genealogy of R comes from the S statistical programming language, created at Bell Labs in the 1970s[4]. As a programming language, R was designed for statisticians to analyze data interactively. R's statistical and academic origins stand in contrast to other programming languages used for data science.

R is an object-based programming language, where code and outputs are stored as objects to be acted upon later. Algebra might store a single value or mathematical expression in a variable; R can hold single or multiple variables,

---

[1]the program that enables you to type commands that your computer will follow to complete a task, such as Terminal on MacOS

[2]https://librarycarpentry.org/lc-shell/01-intro-shell/index.html

[3]https://blog.revolutionanalytics.com/2020/07/the-history-of-r-updated-for-2020.html

[4]https://youtu.be/jk9S3RTAl38

or values, in each object. Where algebra uses an equal sign to denote what a variable is, such as x = 5, R uses <- in the same way. You can read the left-pointing arrow as the word "is." We can use the print() function to display the value of an object when we put the object we want to see inside of the parentheses.

```
x <- 5
y <- x + 2

print(y)
```

```
## [1] 7
```

R lets us work with data interactively through the use of code. When we write code in R, we are usually creating and saving data objects of various classes according to our needs. We can then conduct operations and/or analyses on these data objects in our R **session**(s).

Common classes (types) for these objects include numeric, character (text), and logical (true/false). Objects of a single class are often collected and stored together as vectors. Vectors can in turn be grouped together to make larger data objects you might already be familiar with, including matrices, arrays, or data frames. In this book we focus on data frames.

The data frame structure is central to data analysis because it requires each element of the data frame to have the same length, just like rows in a table. Another way to say this is that each column must be the same length; each column in the data frame must have the same number of rows. This consistent table-like structure is vital for many data science functions. Readers who move on to further data science tasks beyond this book will need to understand data class and structures. Coding errors in R are often traced back to problems with incompatible data structures or inconsistent application of classes.

We can combine multiple values into one object using c() to determine an object's class using class(). Please note that any code preceded by a # functions as a comment because R ignores anything following that character.

```
# numeric vector
numbers <- c(8, 6, 7, 5, 3, 0, 9)

class(numbers)
```

```
## [1] "numeric"
```

```
#logical vector
values <- c(TRUE, TRUE, TRUE, FALSE, FALSE, FALSE, TRUE)

class(values)
```

```
## [1] "logical"
```

Designed explicitly to work with data, R works with many object types. Like many other programming languages, when data scientists find a need for specific applications and groups of related functions, they can create and bundle them into what R calls "packages." A **package** is a group of associated functions equivalent to what in other languages might be called a "library."

The complete list of the thousands and thousands of contributed R packages is on the Comprehensive R Archive Network, or CRAN (https://cran.r-project. org/). These run the gamut from technical to purely fun. Some packages focus on a particular skill (web scraping) or a specific dataset (the Project Gutenberg books). You'll become familiar with a dozen or so packages throughout this book.

2.6 Introducing the Tidyverse

One of the most helpful R packages to become familiar with is the *tidyverse* package, which is a collection of packages[5] usually referred to as the **Tidyverse**. Each of these, listed below, focuses on a different aspect of cleaning or tidying data before it's used or analyzed further. While it is possible to use "base R," meaning the functions that come loaded with R when installing it, many R users prefer to use the Tidyverse because they make common tasks in R easier. The Tidyverse packages all work together, and Posit, PBC staff maintain them.

The "core" Tidyverse packages include

- *ggplot2*, for data visualization
- *dplyr*, for data manipulation
- *tidyr*, for data tidying
- *readr*, for importing data from CSV files
- *purrr*, for functional programming (such as repetitive functions)
- *tibble*, for tibbles, a more straightforward way to create data frames

---

[5]https://www.tidyverse.org/packages/

- *stringr*, for manipulating strings[6]
- *forcats* for factors (a data structure not used in this book)

There are several other additional packages in the Tidyverse, and we will use several of them in this book:

- *httr*, for web APIs
- *jsonlite*, for JSON files
- *readxl*, for .xls and .xlsx files. (not used in this book, but useful for those who use Microsoft Excel frequently)
- *rvest*, for web scraping.
- *xml2*, for working with XML formats

This book will cover the purpose and functions of these packages as they are needed.

2.7 Getting Started with the RStudio IDE

There are many ways to interface with R on your computer, and you can chose the interface that makes the most sense for you. Millions of R users use the graphical user interface provided by RStudio:

The RStudio IDE is a set of integrated tools designed to help you be more productive with R and Python. It includes a console, syntax-highlighting editor that supports direct code execution, and a variety of robust tools for plotting, viewing history, debugging, and managing your workspace[7].

RStudio is also open-source software, which means that the code used to create it is freely available to download, use, and modify. In contrast, other statistical analysis software programs have inaccessible code and require paid subscriptions. Additionally, Posit, PBC supports the continued development of RStudio by dedicating a portion of its engineering team to work only on open-source software projects.

---

[6] https://en.wikipedia.org/wiki/String_(computer_science)
[7] https://posit.co/download/rstudio-desktop/

2.7.1 Install R

The RStudio IDE does not come with R; instead, download the latest version of R for your operating system from CRAN[8]. Follow the download and installation instructions for your operating system to install R.

2.7.2 Install the RStudio IDE

We will use the open-source desktop version of the IDE, which is available as a free download from Posit's website[9]. On the download page, you should select the correct version of the IDE that matches your operating system (OS).

RStudio Desktop 2022.02.2+485 - Release Notes ☑

1. Install R. RStudio requires R 3.3.0+ ☑.

2. Download RStudio Desktop. Recommended for your system:

Requires macOS 10.15+ (64-bit)

All Installers

Linux users may need to import RStudio's public code-signing key ☑ prior to installation, depending on the operating system's security policy.

RStudio requires a 64-bit operating system. If you are on a 32 bit system, you can use an older version of RStudio.

OS	Download	Size	SHA-256
Windows 10/11	⬇ RStudio-2022.02.2-485.exe	177.27 MB	74187a33
macOS 10.15+	⬇ RStudio-2022.02.2-485.dmg	217.09 MB	cda82e98
Ubuntu 18+/Debian 10+	⬇ rstudio-2022.02.2-485-amd64.deb	128.58 MB	508a6e9c
Fedora 19/Red Hat 7	⬇ rstudio-2022.02.2-485-x86_64.rpm	144.66 MB	7400234c

FIGURE 2.1 Select the IDE version that matches your operating system.

After selecting the download button, follow the prompts on your computer to install RStudio.

---

[8]https://cran.r-project.org/
[9]https://posit.co/download/rstudio-desktop/#download

2.7.3 Navigating RStudio

The RStudio IDE brings together all the tools you need to do data science: an editor to write code and text, a console to execute code, access to your computer's terminal, a file explorer, a viewer pane for graphs and visualizations, as well as a version control pane, for those who use Git or Github (more information in the appendix). While it can accommodate many programming languages, the focus of this book will be using RStudio to code in R. Within the ecosystem of R tools, it includes common code libraries and other tools, like spellcheck, which make the work of data science much more manageable.

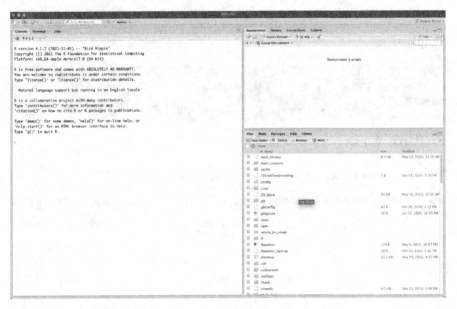

FIGURE 2.2 Open the IDE after download.

RStudio has numerous features, but this book covers only some of those. We'll go over those that are necessary for the tasks at hand. The left-hand pane is called the console, where we can type code directly, or else the IDE will run code within particular files automatically so we can see a log of our code as it executes. Additional tabs are in that pane for the terminal (more information in the appendix). On the top right is the environment pane, where the R objects you create and use in your session are stored. The bottom right is the files pane, where you can navigate through your computer's file directory. Other useful tabs in that pane are Help and Viewer, which shows any graphs or plots you create.

RStudio uses the concept of project files, which group together all the code and dataset files for one project. Every new data science project should start with a new R project in the RStudio IDE. From the **File** menu, select **New Project** and follow the prompts to create a new project. Each project must have a name, which will create a folder of the same name and save all your code and other files within that folder. Naming projects separately keeps project files organized and more easily navigable from a file directory. When you open a project at the start of your work session, the IDE will use the file directory for that project as that session's **working directory**. Any files created will be automatically saved to that same directory or folder, helpful in keeping files organized.

Once you create a file or open one, the console moves to the left bottom, and an editor pane opens in the upper left. To store some R code as a file to access or re-run later, create a new R Script file by going to File > New File > R Script.

FIGURE 2.3 Create a new R Script.

With four panes, the IDE screen looks like this:

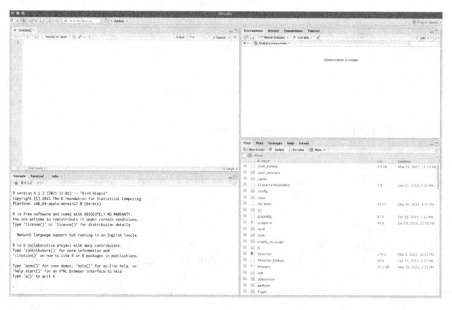

FIGURE 2.4 RStudio IDE with all four panes.

2.8 Packages Needed for this Book

As you progress in your data science journey, you will install more and more R packages. As with any new project you start, begin by installing all the packages you will need to use. Please see the appendix for instructions on installing additional software that these packages depend on to function properly (commonly called "dependencies"). You might see prompts in the console during this process. If you're asked to install other packages, say "yes." If you're asked if you want to compile binaries from source, say "no." Please see the appendix if you encounter error codes during installation.

The R command to install packages is `install.packages()`. Inside the parentheses, you will type the package name, such as *tidyverse*, in quotes:

```
install.packages("tidyverse")
```

For the exercises include in this book, you'll need to install the following packages. We've included a link to information about each package in the table below.

TABLE 2.1 Packages Needed in this Book

Package	Website
tidyverse	https://tidyverse.org
gapminder	https://cran.r-project.org/web/packages/gapminder/index.html
tidytext	https://cran.r-project.org/package=tidytext
jsonlite	https://cran.r-project.org/package=jsonlite
units	https://cran.r-project.org/package=units
rgdal	https://cran.r-project.org/package=rgdal
terra	https://cran.r-project.org/package=terra
sf	https://cran.r-project.org/package=sf
tmap	https://cran.r-project.org/package=tmap
tidycensus	https://walker-data.com/tidycensus/
readr	https://readr.tidyverse.org
textdata	https://cran.r-project.org/package=textdata
tidymodels	https://tidymodels.org
flexdashboard	https://pkgs.rstudio.com/flexdashboard
DT	https://pkgs.rstudio.com/DT
shiny	https://shiny.rstudio.com
Rcpp	https://cran.r-project.org/package=Rcpp
raster	https://cran.r-project.org/package=raster

At the start of all subsequent chapters, you'll notice a code chunk that loads each package into your current session using the `library()` function. Installing a package happens only once, but loading a package must occur each time you open RStudio or start a new R session.

2.9   Viewing Tabular Data in RStudio

Let's read some data into R and get more comfortable with RStudio while exploring the data. We'll use COVID stats for the city of St. Louis that are available at: https://www.stlouis-mo.gov/covid-19/data/#totalsByDate. Scroll down to Totals By Specimen Collection Date and click View Data then save the CSV file.

After the file is saved, we can use the Tidyverse package *readr* and its `read.csv()` function to read the file into R and make it available for us to use. First, we need to load the Tidyverse packages we already installed with the `library()` function.

```
library(tidyverse)

#create an object to store the csv data we read in
stl_covid <- read.csv("City-of-St-Louis-COVID-19-Case-Data.csv")
```

All output needs to be an object, so we created a `stl_covid` object that contains the CSV file we just downloaded. Most COVID datasets are very large, so while we could click on this object in the Environment pane and open it to view the entire file, we could also use a few R functions to get a sense of what this dataset looks like.

If we want to see the entire file, we can use the `view()` command to open up a spreadsheet view in our editor pane. The file is very large, as expected.

```
view(stl_covid)
```

We can also use some built-in base R functions to see snippets of the `stl_covid` dataset. To see the first ten lines, we can use `head()` and `tail()` to see the last ten lines. An additional function is `summary()`, which will display summary statistics for each column in the data frame.

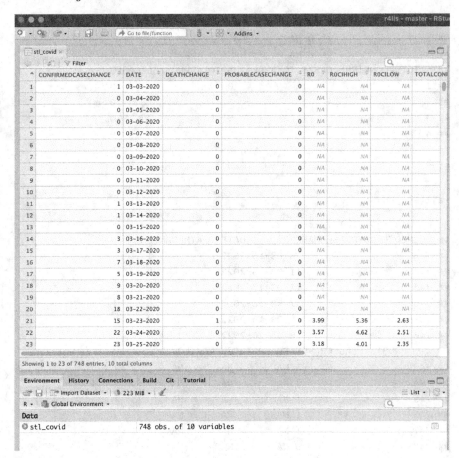

FIGURE 2.5 Viewing the stl_covid object in the IDE.

```
head(stl_covid)
```

```
##   CONFIRMEDCASECHANGE       DATE DEATHCHANGE
## 1                   1 03-03-2020           0
## 2                   0 03-04-2020           0
## 3                   0 03-05-2020           0
## 4                   0 03-06-2020           0
## 5                   0 03-07-2020           0
## 6                   0 03-08-2020           0
##   PROBABLECASECHANGE R0 R0CIHIGH R0CILOW
## 1                  0 NA       NA      NA
## 2                  0 NA       NA      NA
## 3                  0 NA       NA      NA
```

```
## 4                          0 NA          NA          NA
## 5                          0 NA          NA          NA
## 6                          0 NA          NA          NA
##    TOTALCONFIRMEDCASES TOTALDEATHS TOTALPROBABLECASES
## 1                    1           0                  0
## 2                    1           0                  0
## 3                    1           0                  0
## 4                    1           0                  0
## 5                    1           0                  0
## 6                    1           0                  0
```

```
tail(stl_covid)
```

```
##     CONFIRMEDCASECHANGE       DATE DEATHCHANGE
## 743                   8 03-15-2022           1
## 744                  11 03-16-2022           0
## 745                  11 03-17-2022           0
## 746                   7 03-18-2022           0
## 747                   5 03-19-2022           0
## 748                   0 03-20-2022           0
##     PROBABLECASECHANGE R0 R0CIHIGH R0CILOW
## 743                  5 NA       NA      NA
## 744                  3 NA       NA      NA
## 745                  0 NA       NA      NA
## 746                  1 NA       NA      NA
## 747                  1 NA       NA      NA
## 748                  0 NA       NA      NA
##     TOTALCONFIRMEDCASES TOTALDEATHS TOTALPROBABLECASES
## 743               45378         746               7504
## 744               45389         746               7507
## 745               45400         746               7507
## 746               45407         746               7508
## 747               45412         746               7509
## 748               45412         746               7509
```

```
summary(stl_covid)
```

```
##  CONFIRMEDCASECHANGE     DATE
##  Min.   :  0.0       Length:748
##  1st Qu.: 21.0       Class :character
##  Median : 35.0       Mode  :character
##  Mean   : 60.7
##  3rd Qu.: 66.0
##  Max.   :735.0
```

```
##
##    DEATHCHANGE        PROBABLECASECHANGE        R0
##   Min.   : 0.000     Min.   : -2          Min.   :0.56
##   1st Qu.: 0.000     1st Qu.:  0          1st Qu.:0.89
##   Median : 0.000     Median :  5          Median :0.99
##   Mean   : 0.997     Mean   : 10          Mean   :1.04
##   3rd Qu.: 2.000     3rd Qu.: 12          3rd Qu.:1.14
##   Max.   :12.000     Max.   :241          Max.   :3.99
##                                           NA's   :34
##     R0CIHIGH          R0CILOW       TOTALCONFIRMEDCASES
##   Min.   :0.61     Min.   :0.51     Min.   :    1
##   1st Qu.:1.00     1st Qu.:0.76     1st Qu.: 6541
##   Median :1.11     Median :0.89     Median :20429
##   Mean   :1.16     Mean   :0.91     Mean   :18537
##   3rd Qu.:1.28     3rd Qu.:1.00     3rd Qu.:26597
##   Max.   :5.36     Max.   :2.63     Max.   :45412
##   NA's   :34       NA's   :34
##   TOTALDEATHS    TOTALPROBABLECASES
##   Min.   :  0    Min.   :   0
##   1st Qu.:214    1st Qu.:  97
##   Median :438    Median :1724
##   Mean   :396    Mean   :2125
##   3rd Qu.:577    3rd Qu.:3401
##   Max.   :746    Max.   :7509
##
```

While this dataset originated as a CSV file, there are specific R packages for reading in Microsoft Excel (*readxl*) and Google Sheets (*googlesheets4*). This book works only with CSV files, but know that if you often work with those proprietary formats, other packages exist to help you with those if you don't want to convert them to CSV files first.

2.10 Summary

This chapter took you from no experience coding in R to interacting with data in the RStudio IDE using R functions. R is an object-oriented programming language used within RStudio's graphical user interface alongside several popular code packages, such as the Tidyverse. New users must install R and RStudio before learning the various features the IDE offers for data scientists.

There are several ways to view data in RStudio, whether viewing the entire dataset file or using R functions to see snippets of the dataset within the console.

2.11   Further Practice

- Read in a CSV file of your own and run the same summary functions: `head()`, `tail()`, `summary()`
- Install *janeaustenR* for use in Chapter 6

2.12   Additional Resources

- *Hands-on programming with R*: https://rstudio-education.github.io/hopr/
- RStudio IDE, Base R, & data import (*readr*) cheatsheets: https://posit.co/resources/cheatsheets/
- "Getting Started with R and RStudio": https://moderndive.netlify.com/1-getting-started.html
- An Introduction to R: https://cran.r-project.org/doc/manuals/R-intro.html

3

Tidying Data with dplyr

3.1 Learning Objectives

1. Use the IDE to load the *dplyr* package.
2. Identify data elements in RStudio's IDE that need to be changed.
3. Summarize the most common functions *dplyr* is used for.
4. Use *dplyr* functions to normalize fields in a dataset.

3.2 Terms You'll Learn

- API

3.3 Scenario

You need data on unemployment in the city of St. Louis, and the first step to creating visualizations related to unemployment requires you to read the data and tidy it. You'd like to target your outreach to areas of low unemployment, so you will need to prepare data to use in determining those. Occupations with the highest employment would be helpful to target training for job seekers for jobs that are in demand.

DOI: 10.1201/9781003218012-3

3.4   Packages & Datasets Needed

```
library(tidyverse)
library(units)
library(sf)
library(tmap)
library(tidycensus)
```

3.5   Introduction

This chapter is focused on Census data and learning data tidying functions to create an unemployment dataset for use in subsequent chapters. We are aided in this endeavor by the *tidycensus* package[1], which interfaces with US Census datasets and returns data that are ready to work with Tidyverse packages. *Tidycensus* lets us access Census data for many communities, St. Louis included. The Census contains data about employment, occupation, gender, and location.

3.6   Getting Started with U.S. Census Data

Census data is available from the Census **API**[2]. An API, or application programming interface, allows our computer to access the computer(s) storing the census data. APIs enable computers to talk to each other; they are a valuable tool for data scientists who want to get a dataset directly from the source. Many data sources provide API access to their databases, which we will visit again in chapter 7.

---

[1]https://walker-data.com/tidycensus/
[2]https://en.wikipedia.org/wiki/API

3.6.1 Census Prerequisites

Before using *tidycensus* to query the Census database, each user must have a unique identifier: an API key. This unique authorization code from the Census website allows you to access Census data[3].

1. Create a Census API key

If you're following along and entering this code into your R console, sign up for your own census data key and replace '"your-key-here" with your own API key.

```
census_api_key("your-key-here")
```

2. Get FIPS codes

We are limiting our analysis to the city of St. Louis and need to restrict our data to that area. To do that, we'll use the Federal Information Processing Series (FIPS) Codes. Thankfully, `fips_codes` are already part of *tidycensus*.

```
head(fips_codes)
```

```
##    state state_code state_name county_code
## 1     AL         01    Alabama         001
## 2     AL         01    Alabama         003
## 3     AL         01    Alabama         005
## 4     AL         01    Alabama         007
## 5     AL         01    Alabama         009
## 6     AL         01    Alabama         011
##              county
## 1 Autauga County
## 2 Baldwin County
## 3 Barbour County
## 4    Bibb County
## 5  Blount County
## 6 Bullock County
```

When combined with the state, each county has a code that allows us to query the Census database for only the geographic area of interest, like St. Louis.

---

[3]http://api.census.gov/data/key_signup.html

3.6.2 Census Variables

The Census collects a lot of data about the U.S. population, but we don't
need all that data! To narrow our scope to the most applicable data, we must
select the Census report year, type, and metadata fields (variables) we want
to analyze. The American Community Survey[4] will provide the most valuable
data for our analysis.

1. Review all Census variables

We'll use load_variables() to review the 2019 ACS 5-year survey data variables.

```
var_2019 <- load_variables(2019, "acs5")
var_2019
```

```
## # A tibble: 27,040 x 4
##     name       label                    concept geogr~1
##     <chr>      <chr>                    <chr>   <chr>
##  1 B01001_001 Estimate!!Total:          SEX BY~ block ~
##  2 B01001_002 Estimate!!Total:!!Male:   SEX BY~ block ~
##  3 B01001_003 Estimate!!Total:!!Male:!~ SEX BY~ block ~
##  4 B01001_004 Estimate!!Total:!!Male:!~ SEX BY~ block ~
##  5 B01001_005 Estimate!!Total:!!Male:!~ SEX BY~ block ~
##  6 B01001_006 Estimate!!Total:!!Male:!~ SEX BY~ block ~
##  7 B01001_007 Estimate!!Total:!!Male:!~ SEX BY~ block ~
##  8 B01001_008 Estimate!!Total:!!Male:!~ SEX BY~ block ~
##  9 B01001_009 Estimate!!Total:!!Male:!~ SEX BY~ block ~
## 10 B01001_010 Estimate!!Total:!!Male:!~ SEX BY~ block ~
## # ... with 27,030 more rows, and abbreviated variable
## #   name 1: geography
```

2. Create new objects for variables

Having pulled in the FIPS codes that allow us to identify data from St. Louis
and the variable names from the 2019 ACS, we can now create a new object
that contains only the data we want:

- Survey: 5-year ACS
- Year: 2019
- Locations: St. Louis County, Missouri

---

[4]https://www.census.gov/programs-surveys/acs

- Total population: B23025_001
- Population not in the labor force (unemployed): B23025_007

One Base R function that we'll rely on for this code is c(), which concatenates strings (numbers or text) into one value. We'll concatenate the two variables we're interested in: total population and the number of unemployed. The function get_acs() passes the metadata requirements to the Census database, returning the data we need for each Census tract. We're interested in all the variables and want to see them spread out across the columns, so we will use the output = "wide" setting to adjust the output.

```r
data <- get_acs(
  geography = "tract",
  variables = c(
    total_pop = "B23001_001",
    unemployed = "B23025_007"
  ),
  state = "MO",
  county = "510",
  year = 2019,
  output = "wide"
)
```

```
## Getting data from the 2015-2019 5-year ACS
```

```r
data
```

```
## # A tibble: 106 x 6
##    GEOID       NAME        total~1 total~2 unemp~3 unemp~4
##    <chr>       <chr>         <dbl>   <dbl>   <dbl>   <dbl>
## 1  29510124200 Census ~       2536     326     668     194
## 2  29510124300 Census ~       3043     305     649     142
## 3  29510125500 Census ~       3881     328     791     177
## 4  29510102100 Census ~       2300     184     545     107
## 5  29510102400 Census ~       2086     166     570     152
## 6  29510103800 Census ~       3269     240     898     131
## 7  29510104200 Census ~       3000     261     869     218
## 8  29510105500 Census ~       2265     388     964     215
## 9  29510106500 Census ~       2275     376    1135     265
## 10 29510107500 Census ~       1730     312     904     220
## # ... with 96 more rows, and abbreviated variable
## #   names 1: total_popE, 2: total_popM,
## #   3: unemployedE, 4: unemployedM
```

3.7   Tidy Data Tools from *dplyr*

The data we've pulled is the total population and the number of unemployed, but that's not what we need to know. We need an unemployment rate; from there we can determine where the areas of highest and lowest unemployment are alongside occupation data. To do this, we must tidy and modify the *tidycensus* data we have.

The *dplyr* package within the Tidyverse contains a constellation of functions designed for data modification. Some of the actions we'll need to perform are

- renaming columns
- creating a new column for the unemployment rate, which involves performing a mathematical operation on other columns
- combine columns
- sort column values
- combine several functions sequentially
- choose specific columns or rows within the table
- see a snapshot of a dataset
- group data by column value
- filter a subset of a table
- combine datasets based on common column values

3.8   Getting Started with *dplyr* Functions

One of the formative concepts of the Tidyverse, which we will rely upon heavily through the remainder of the book, is the use of the pipe: %>%. This operator can be read within a code chunk as "then"; it allows us to call *dplyr* functions sequentially to make our code more readable. You will often see code from the Tidyverse written in an "object [then] function" syntax pattern.

3.8.1   Create Unemployment Rate

We'll use the "object [then] function" pattern to create a new variable and column for the unemployment rate. The ACS doesn't provide an unemployment rate, so we must calculate it from the columns we have, total population and population unemployed. Our task here is two fold: 1) create a new column and 2) populate each row in the column with its calculated unemployment rate. This is the number unemployed divided by the total population:

```r
unemployment_data <- data %>%
  mutate(
    unemployment_rate =
      as.numeric(unemployedE) / as.numeric(total_popE)
  )
```

In plain English, the code above says "create a new object called unemployment_data, which takes the data object and then makes a new column in it called unemployment_rate; fill the rows in that new column with the value of the number unemployed divided by the total population."

3.8.2 Save Unemployment Data

Before going any further, we will save the unemployment_data object to a CSV file in the data/ sub-directory, or folder, using the write_csv() function from *readr*:

```r
write_csv(unemployment_data, "data/unemployment_data.csv")
```

3.8.3 Find the Areas with the Highest Unemployment

Getting back to *dplyr*, we need to figure out where the areas with the highest levels of unemployment are. We'll use arrange() to sort the dataset by unemployment rate in descending order and then look at only the top 10 locations in the dataset.

```r
# `arrange()` defaults to ascending sort order
high_unemploy <- unemployment_data %>%
  arrange(desc(unemployment_rate)) %>%
  head(n = 10)

high_unemploy
```

```
## # A tibble: 10 x 7
##     GEOID  NAME   total~1 total~2 unemp~3 unemp~4 unemp~5
##     <chr>  <chr>   <dbl>   <dbl>   <dbl>   <dbl>   <dbl>
##  1 29510~ Cens~    1572     337    1019     262   0.648
```

```
## 2 29510~ Cens~    1448    169     866    166   0.598
## 3 29510~ Cens~    2360    345    1376    252   0.583
## 4 29510~ Cens~    1299    220     743    137   0.572
## 5 29510~ Cens~    1333    172     733    126   0.550
## 6 29510~ Cens~    2163    287    1174    197   0.543
## 7 29510~ Cens~    3721    345    2011    271   0.540
## 8 29510~ Cens~    1613    198     858    197   0.532
## 9 29510~ Cens~    1769    218     930    182   0.526
## 10 29510~ Cens~   1730    312     904    220   0.523
## # ... with abbreviated variable names 1: total_popE,
## #    2: total_popM, 3: unemployedE, 4: unemployedM,
## #    5: unemployment_rate
```

3.8.4   Find the Areas with the Lowest Unemployment

Using the same arrange() function, but using the default ascending sort, we'll create a new object of only the values with the lowest unemployment rate.

```
low_unemploy <- unemployment_data %>%
  arrange(unemployment_rate) %>%
  head(n = 10)


low_unemploy
```

```
## # A tibble: 10 x 7
##     GEOID  NAME   total~1 total~2 unemp~3 unemp~4 unemp~5
##     <chr>  <chr>   <dbl>   <dbl>   <dbl>   <dbl>   <dbl>
## 1  29510~ Cens~    2362    227     417    133   0.177
## 2  29510~ Cens~    2617    262     511    157   0.195
## 3  29510~ Cens~    3556    311     720    177   0.202
## 4  29510~ Cens~    3881    328     791    177   0.204
## 5  29510~ Cens~    1401    139     286     83   0.204
## 6  29510~ Cens~    3561    442     734    223   0.206
## 7  29510~ Cens~    3420    298     712    157   0.208
## 8  29510~ Cens~    1612    155     336     77   0.208
## 9  29510~ Cens~    2262    206     480    117   0.212
## 10 29510~ Cens~    3043    305     649    142   0.213
## # ... with abbreviated variable names 1: total_popE,
## #    2: total_popM, 3: unemployedE, 4: unemployedM,
## #    5: unemployment_rate
```

3.9 Occupation Data

Now that we have the top ten Census block groups with the highest and lowest unemployment rates, let's see the occupations in those Census blocks. We will select the full table of data instead of only specific variables.

3.9.1 Occupation Data for the City of St. Louis

When we use the `get_acs()` function to pull that survey's data from the census website, we need to specify which particular table of data we need. "Tract" is the unit of geographical measurement we need; we'll include the year, county/state, and the code for the table. A Census tract is a usually permanent subdivision of a county with about 4,000 people that reside within its bounds[5]. We'll use the *dplyr* function `glimpse()` to see a list of column names and a snippet of the values for each. This function is handy to see an object at a glance to understand what it contains.

```
occupation_data <- get_acs(
  geography = "tract",
  state = "MO",
  county = "510",
  year = 2019,
  table = "C24010"
)
```

```
## Getting data from the 2015-2019 5-year ACS
```

```
glimpse(occupation_data)
```

```
## Rows: 7,738
## Columns: 5
## $ GEOID    <chr> "29510101100", "29510101100", "29510~
## $ NAME     <chr> "Census Tract 1011, St. Louis city, ~
## $ variable <chr> "C24010_001", "C24010_002", "C24010_~
## $ estimate <dbl> 1338, 744, 156, 89, 73, 16, 34, 8, 1~
## $ moe      <dbl> 136, 106, 70, 56, 50, 19, 28, 13, 18~
```

---

[5]https://www.census.gov/programs-surveys/geography/about/glossary.html#par_text
image__13

The GEOID variable, which contains the Census tract ID, is present in both the occupation data and the employment data.

3.9.2   Save Occupation Data

We'll save the occupation data to a CSV file so we can use it later.

```
write_csv(occupation_data, "data/occupation_data.csv")
```

3.9.3   Select Occupations with the Highest and Lowest Unemployment

We will need to use `filter()` to get the occupations with the highest and lowest unemployment rates and then use `arrange()` again to sort the results. Our challenge here is that we don't want all the occupation data; we only need the occupations for the areas we already identified in the `low_unemploy` and `high_unemploy`objects we created in the previous section. We'll need to use the `%in%` operator, which lets us select only specific rows that match both datasets.

To find the occupations (jobs) with the lowest unemployment, we want to take the `occupation_data` object and filter out the rows whose GEOID matches the GEOID field in the `low_unemploy` object. Then we want to arrange them in descending order by the `estimate` column.

```
low_unemploy_jobs <- occupation_data %>%
  filter(GEOID %in% low_unemploy$GEOID) %>%
  arrange(desc(estimate))

glimpse(low_unemploy_jobs)
```

```
## Rows: 730
## Columns: 5
## $ GEOID    <chr> "29510125500", "29510114101", "29510~
## $ NAME     <chr> "Census Tract 1255, St. Louis city, ~
## $ variable <chr> "C24010_001", "C24010_001", "C24010_~
## $ estimate <dbl> 2894, 2637, 2598, 2440, 2188, 2050, ~
## $ moe      <dbl> 340, 317, 259, 385, 240, 230, 175, 2~
```

We'll perform the same functions for jobs with the highest unemployment but compare rows against the `high_unemploy` object.

```
high_unemploy_jobs <- occupation_data %>%
  filter(GEOID %in% high_unemploy$GEOID) %>%
  arrange(desc(estimate))

glimpse(high_unemploy_jobs)
```

```
## Rows: 730
## Columns: 5
## $ GEOID    <chr> "29510119300", "29510106700", "29510~
## $ NAME     <chr> "Census Tract 1193, St. Louis city, ~
## $ variable <chr> "C24010_001", "C24010_001", "C24010_~
## $ estimate <dbl> 1619, 952, 848, 794, 771, 714, 700, ~
## $ moe      <dbl> 221, 263, 146, 225, 147, 147, 145, 2~
```

3.9.4 Combine Occupation Names with Unemployment Data

The high and low unemployment datasets include the occupation in the
variable column, but only as a code. At the beginning of this chapter, we
read the list of variables and saved it to the var_2019 object. We need to combine these two datasets to bring only the variables we need from var_2019 to
high_unemploy_jobs and low_unemploy_jobs.

dplyr has several "join" functions that allow you to combine and filter data in
one step. This book will utilize:

- "left_join()"[6]

  - Adds columns from the second dataset to the first dataset

- "anti_join()"[7]

  - Returns all rows from the first dataset that **do not** have a match in the
    second dataset

To combine the variable names with the code that matches our low unemployment occupations, we'll take our low_unemploy_jobs dataset and use
left_join() to add fields from var_2019. We need to indicate which column
we should join the dataset by, which is another way of identifying which column is present in both datasets. The column appearing in both datasets will
match the rest of the rows and columns together.

```
low_unemploy_jobs_join <- low_unemploy_jobs %>%
  left_join(var_2019, by = c("variable" = "name"))

glimpse(low_unemploy_jobs_join)
```

---

[6]https://dplyr.tidyverse.org/reference/mutate-joins.html
[7]https://dplyr.tidyverse.org/reference/filter-joins.html

```
## Rows: 730
## Columns: 8
## $ GEOID     <chr> "29510125500", "29510114101", "2951~
## $ NAME      <chr> "Census Tract 1255, St. Louis city,~
## $ variable  <chr> "C24010_001", "C24010_001", "C24010~
## $ estimate  <dbl> 2894, 2637, 2598, 2440, 2188, 2050,~
## $ moe       <dbl> 340, 317, 259, 385, 240, 230, 175, ~
## $ label     <chr> "Estimate!!Total:", "Estimate!!Tota~
## $ concept   <chr> "SEX BY OCCUPATION FOR THE CIVILIAN~
## $ geography <chr> "block group", "block group", "bloc~
```

When we look at the combined dataset, we can see only one variable (name) column, but otherwise, the left_join() added all the columns from var_2019 to low_unemploy_jobs.

3.9.5   Group Occupations by Gender

Census data includes binary gender classifications (male/female) for each occupation. We can de-duplicate the dataset by separating the rows by gender.

We'll use several new *dplyr* functions to refine the dataset. To group variables with the same values, group_by() lets us specify which column name to use. The pipe comes in handy as we work several sequential steps on the same low_unemploy_jobs_join dataset. First, we'll group the values by label. Then we'll use summarize()[8], which is a function that will create a new table of summarized data containing the grouped labels and the column called total that will contain the total estimate column. We can then arrange the values in descending order. The last function we will pipe into this object is filter(), which we'll use to exclude gross aggregate estimates that are over 10,000. Excluded values include the total estimate and total female estimate rows because we're interested in specific occupation data.

```
low_unemploy_jobs_ct <- low_unemploy_jobs_join %>%
  group_by(label) %>%
  summarize(total = sum(estimate)) %>%
  arrange(desc(total)) %>%
  filter(total < 10000)
```

---

[8]https://dplyr.tidyverse.org/reference/summarise.html

3.9.6   Occupations with the Lowest Unemployment for Men

Because the occupation rankings differ for men and women, we need to separate the data into two datasets by gender. To do that, we'll utilize the *stringr* package. We can use the `str_detect()` function to `filter()` rows that match the character string we specify. In this case, we want to filter the label column for the value of "Male." Then we can arrange the results in descending order and look at the top 10 occupations with the lowest unemployment rate for men.

```
low_unemploy_jobs_male <- low_unemploy_jobs_ct %>%
  filter(str_detect(label, "Male")) %>%
  arrange(desc(total)) %>%
  head(n = 10)


low_unemploy_jobs_male
```

```
## # A tibble: 10 x 2
##    label                                              total
##    <chr>                                              <dbl>
##  1 Estimate!!Total:!!Male:!!Management, business~      5128
##  2 Estimate!!Total:!!Male:!!Management, business~      2072
##  3 Estimate!!Total:!!Male:!!Service occupations:       1586
##  4 Estimate!!Total:!!Male:!!Sales and office occ~      1494
##  5 Estimate!!Total:!!Male:!!Management, business~      1424
##  6 Estimate!!Total:!!Male:!!Management, business~      1271
##  7 Estimate!!Total:!!Male:!!Management, business~      1219
##  8 Estimate!!Total:!!Male:!!Production, transpor~      1164
##  9 Estimate!!Total:!!Male:!!Sales and office occ~       868
## 10 Estimate!!Total:!!Male:!!Management, business~       867
```

3.9.7   Occupations with the Lowest Unemployment for Women

To create a new dataset showing occupations with the lowest unemployment for women, we will follow the same steps except for searching the label column for "Female."

```
low_unemploy_jobs_female <- low_unemploy_jobs_ct %>% # clean
  filter(str_detect(label, "Female")) %>%
  arrange(desc(total)) %>%
  head(n = 10)


low_unemploy_jobs_female
```

```
## # A tibble: 10 x 2
##    label                                            total
##    <chr>                                            <dbl>
##  1 Estimate!!Total:!!Female:!!Management, busine~    5648
##  2 Estimate!!Total:!!Female:!!Sales and office o~    2549
##  3 Estimate!!Total:!!Female:!!Management, busine~    2095
##  4 Estimate!!Total:!!Female:!!Management, busine~    1737
##  5 Estimate!!Total:!!Female:!!Service occupation~    1677
##  6 Estimate!!Total:!!Female:!!Sales and office o~    1360
##  7 Estimate!!Total:!!Female:!!Management, busine~    1225
##  8 Estimate!!Total:!!Female:!!Sales and office o~    1189
##  9 Estimate!!Total:!!Female:!!Management, busine~    1103
## 10 Estimate!!Total:!!Female:!!Management, busine~     870
```

The occupation with the lowest unemployment is the same for men and women: management, business, science, and the arts. The remaining nine lowest unemployment occupations differ between genders, though.

3.10   Clean Up Metadata

As we can see from our glimpse of the occupations for women with the lowest unemployment, the occupation labels from the Census are tough to parse. There are many exclamation points and colons, and several words repeat in addition to the occupation category itself. Knowing that we want to present this information in a professional context, we will have to spend some time simplifying and relabeling each occupation.

3.10.1   Simplify and Relabel Male Occupations

Cleaning up column values is not an unusual tidying task. In this instance, the irregularities of these values and the small number (ten) of each lend themselves to creating a new object with clean names that we can join with the dataset to provide comprehensible occupation names.

We'll use *tibble*[9], another core Tidyverse package to create a tibble (a tidy table) of occupation labels using `tribble()`[10]. When using `tribble()`, we first include the names of the two columns in our tibble: `label`, and `male_jobs`.

---

[9]https://tibble.tidyverse.org/index.html
[10]https://tibble.tidyverse.org/reference/tribble.html

Then we proceed to include each value in the tibble we want to create, column by column and row by row. We will take the existing values for the `label` column from the Census and create a clean version of each occupation in a new column called `male_jobs`.

```
clean_labels_m <- tribble(
  ~label, ~male_jobs,
  "Estimate!!Total:!!Male:!!Management, business, science, and arts
    occupations:",
  "Management, business, science, & arts",
  "Estimate!!Total:!!Male:!!Management, business, science, and arts
    occupations:!!Management, business, and financial occupations:",
  "Business & financial operations",
  "Estimate!!Total:!!Male:!!Service occupations:",
  "Service occupations",
  "Estimate!!Total:!!Male:!!Sales and office occupations:",
  "Sales & office",
  "Estimate!!Total:!!Male:!!Management, business, science, and arts
    occupations:!!Computer, engineering, and science occupations:",
  "Computer, engineering, & science",
  "Estimate!!Total:!!Male:!!Management, business, science, and arts
    occupations:!!Management, business, and financial
    occupations:!!Management occupations",
  "Management",
  "Estimate!!Total:!!Male:!!Management, business, science, and arts
    occupations:!!Education, legal, community service, arts, and
    media occupations:",
  "Education, legal, community service, arts, & media",
  "Estimate!!Total:!!Male:!!Production, transportation, and material
    moving occupations:",
  "Production, transportation, & material moving",
  "Estimate!!Total:!!Male:!!Sales and office occupations:!!Sales and
    related occupations",
  "Sales",
  "Estimate!!Total:!!Male:!!Management, business, science, and arts
    occupations:!!Computer, engineering, and science
    occupations:!!Computer and mathematical occupations",
  "Computer & mathematical"
)
```

In order to see the beginning of our second clean tidy table, we'll run `glimpse()`.

```
glimpse(clean_labels_m)
```

```
## Rows: 10
## Columns: 2
## $ label      <chr> "Estimate!!Total:!!Male:!!Managemen~
## $ male_jobs <chr> "Management, business, science, & a~
```

3.10.2 Update Occupation Labels for Male Unemployment Data

Now that we have a table of clean occupation labels, we can use another
left_join() to combine clean_labels_m with low_unemploy_jobs_male.

```
low_unemploy_jobs_malec <- left_join(
  low_unemploy_jobs_male,
  clean_labels_m,
  by = "label"
)
```

```
glimpse(low_unemploy_jobs_malec)
```

```
Rows: 10
Columns: 3
$ label      <chr> "Estimate!!Total:!!Male:!!Management, business, science, and a...
$ total      <dbl> 5128, 2072, 1586, 1494, 1424, 1271, 1219, 1164, 868, 867
$ male_jobs <chr> "Management, business, science, & "arts", "Business & financial...
```

3.10.3 Simplify and Relabel Female Occupations

We'll repeat the same occupation label cleaning steps that we did with male
occupations with the female occupations using tribble(). Our column names
are label and female_jobs, and then we will enter the existing labels and their
clean counterparts.

```
clean_labels_f <- tribble(
  ~label, ~female_jobs,
  "Estimate!!Total:!!Female:!!Management, business, science, and
    arts occupations:",
  "Management, business, science, & arts",
  "Estimate!!Total:!!Female:!!Service occupations:",
  "Service occupations",
  "Estimate!!Total:!!Female:!!Sales and office occupations:!!Sales
    and related occupations",
  "Sales",
```

```
  "Estimate!!Total:!!Female:!!Sales and office occupations:!!Office
    and administrative support occupations",
  "Office & administrative support",
  "Estimate!!Total:!!Female:!!Sales and office occupations:",
  "Sales & office",
  "Estimate!!Total:!!Female:!!Management, business, science, and
    arts occupations:!!Management, business, and financial
    occupations:!!Management occupations",
  "Management",
  "Estimate!!Total:!!Female:!!Management, business, science, and
    arts occupations:!!Management, business, and financial
    occupations:!!Business and financial operations occupations",
  "Business & financial operations",
  "Estimate!!Total:!!Female:!!Management, business, science, and
    arts occupations:!!Management, business, and financial
    occupations:",
  "Management, business, & financial",
  "Estimate!!Total:!!Female:!!Management, business, science, and
    arts occupations:!!Healthcare practitioners and technical
    occupations:",
  "Healthcare practitioners & technical",
  "Estimate!!Total:!!Female:!!Management, business, science, and
    arts occupations:!!!Education, legal, community service, arts,
    and media occupations:",
  "Education, legal, community service, arts, & media"
)
```

In order to see the beginning of our clean tidy table, we'll run `glimpse()`.

```
glimpse(clean_labels_f)
```

```
## Rows: 10
## Columns: 2
## $ label       <chr> "Estimate!!Total:!!Female:!!Manag~
## $ female_jobs <chr> "Management, business, science, &~
```

3.10.4 Update Occupation Labels for Female Unemployment Data

Again, we'll use the previous tool of `left_join()` to combine the clean labels with the list of female unemployment data we have.

```
low_unemploy_jobs_femalec <- left_join(
  low_unemploy_jobs_female,
  clean_labels_f,
  by = "label"
)
```

3.11 Create a CSV of Unemployment Data

All of the data tidying work we've done in this chapter needs to be saved as new files to use in later chapters. We'll use *readr* again to save the male and female unemployment data to individual CSV files.

3.11.1 1. Male Unemployment

```
write_csv(low_unemploy_jobs_malec, "data/male-low-unemployment.csv")
```

3.11.2 2. Female Unemployment

```
write_csv(low_unemploy_jobs_femalec, "data/female-low-unemployment.csv")
```

3.12 Summary

The *dplyr* package is a foundational Tidyverse package. We used it in this chapter to modify our census data into tables that we can use in later chapters to analyze and plot in R. We've sorted, renamed, grouped, joined, and filtered St. Louis Census data. Other functions have allowed us to create new columns with new data, such as `mutate()` and `summarize()`. Data rarely arrives in a state perfectly ready for analysis in the real world. Messy data makes learning data cleaning functions essential for work in data science. We also learned to use the central Tidyverse operator: the pipe. With the pipe, we can stack functions one after the other to manipulate our data efficiently and quickly. Using *dplyr* functions with the pipe presents a stark contrast to years of working with Excel files that have to be manually modified repeatedly in (hopefully) the same way.

3.13 Further Practice

- Use the Census API to import a different ACS year for St. Louis and adjust the column names and variables as appropriate.

3.14 Additional Resources

- Data transformation (*dplyr*) cheatsheet: https://posit.co/resources/cheatsheets/
- *R for Data Science*: https://r4ds.had.co.nz/

4

Visualizing Your Project with ggplot2

4.1 Learning Objectives

1. Describe various geom functions in *ggplot2* for making plots.
2. Use appropriate code to load the *ggplot2* library and other relevant libraries into RStudio.
3. Generate basic column plots using the relevant geom functions in *ggplot2*.
4. Apply the `aes()` function to customize plots.
5. Apply legends and labels to plots.

4.2 Terms You'll Learn

- Aesthetic mappings
- Geometric object
- Coordinate system
- Polar coordinate system
- Facet
- Scales
- Theme

4.3 Scenario

You were asked by your library to generate some data visualizations about the demographic profile of St. Louis City. Specifically, you need data visualizations

DOI: 10.1201/9781003218012-4

of the top ten census block groups and occupations with the highest and lowest unemployment rates.

4.4   Packages & Datasets Needed

```
library(tidyverse)
library(tidycensus)
library(gapminder)
library(readr)
```

4.5   Introduction to *ggplot2*

The *ggplot2* package creates data visualizations such as histograms and plots. Its syntax comes from Leland Wilkinson's *Grammar of Graphics* (Wilkinson, 2005), which makes layers of graphics by mapping aesthetic attributes to geometric objects (Navarro and Pedersen, 2020). Aesthetic attributes are things like size, color, and shape. A **geometric object** can be a plot that could be a bar, line, or point plot. For this exercise, we will focus on creating bar plots because we will be working with one discrete variable, the Census tracts, and one continuous variable, the unemployment rates. Data scientists refer to such data visualizations as "plots" regardless of the type of data visualization.

4.6   Components of a plot

A typical plot contains data, a coordinate system, and a geometry. The bare minimum code to create a plot is: ggplot(data = DATA, mapping =

aes(MAPPINGS)) + geom_function(). The gg in ggplot stands for the "Grammar of Graphics" that was introduced in the previous section.

The function ggplot() calls the package, and in the parenthesis you apply **aesthetic mappings** to your data. For example, aesthetic mappings indicate which variable is on the x-axis and which is on the y-axis. You can also apply different stylistic mappings to the plot by plotting the points by size, shape, or color. Then, you add a layer using the geom() function, which, for example, could be a histogram (geom_hist()), scatterplot (geom_point()), or a line plot (geom_line()). This list is not exhaustive; the *ggplot2* cheat sheet[1] lists all of the varieties of plots that are available.

You can further customize your plot by specifying the statistical transformation (also known as stat, see section 3.7 in *R for Data Science*[2]) and position of how you display your values (simply known as position). You can also add a coordinate system, facets, scales, and themes.

4.7 Coordinate Systems

You plot your data on a **coordinate system**, also known as a coordinate reference system. Those terms will be used interchangeably in this book. Coordinate system options in *ggplot2* include a cartesian coordinate system, a **polar coordinate system**, and a spatial coordinate system.

Cartesian coordinate system: A two-dimensional coordinate system based on a horizontal x-axis, a vertical y-axis, and diagonal z-axis. You can also flip the x and y axes, plot the coordinates on a fixed ratio, and transform coordinates.

Polar coordinate system: A two-dimensional coordinate system that consists of a reference point and an angle from a reference direction

Spatial coordinate system: A spatial coordinate system comprises lines of latitude which run parallel to the equator and longitude, which runs parallel to the prime meridian. We will go more into spatial coordinate systems in Chapter 6, using the *tmap* package.

---

[1] https://posit.co/resources/cheatsheets/
[2] https://r4ds.had.co.nz/data-visualisation.html

4.8   Unemployment plots

We will be using the `male_low_unemployment`, `female_low_employment`, and un-
employment_data CSV files that we created in Chapter 3. Make sure that you
have these data imported in RStudio.

```
male_low_unemployment <- read_csv("data/male-low-unemployment.csv")
female_low_unemployment <- read_csv("data/female-low-unemployment.csv")
unemployment_data <- read_csv("data/unemployment_data.csv")
```

We'll also recreate the high and low unemployment objects we used in Chapter
3.

```
high_unemploy <- unemployment_data %>%
  arrange(desc(unemployment_rate)) %>%
  head(n = 10)

low_unemploy <- unemployment_data %>%
  arrange(unemployment_rate) %>%
  head(n = 10)
```

4.8.1   Plotting the Census Block Groups by Unemployment Rate with `geom_col()`

You use `geom_col()` for a continuous and discrete variable. To add a layer
to your plot, you will need to use + and then call your function afterwards.
First, we will plot the ten Census block groups with the highest and lowest
unemployment rates:

```
# Ten Census block groups with the highest unemployment rates
high_unemp_plot <- high_unemploy %>%
  ggplot(aes(
    x = unemployment_rate,
    y = reorder(NAME, unemployment_rate)
  )) +
  geom_col()

high_unemp_plot
```

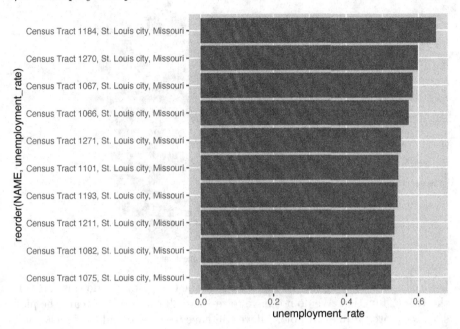

Let's break down this code. We are creating a variable called high_unemp_plot which uses the high_unemploy data which is passed to the ggplot() function via the pipe (%>%) operator. We then call the ggplot() function in which the unemployment rate will be on the x-axis whlie the Census tract name is on the y-axis. The reorder function will reorder the Census tracts based on the unemployment rate. As a result, the Census tract with the highest unemployment rate will be on top of the y-axis while the Census tract with the lowest unemployment rate will be on the bottom of the y-axis. The same method is applied to create the low_unemp_plot to indicate the Census tracts with the lowest rates of unemployment.

```
# Ten areas with the lowest unemployment rates
low_unemp_plot <- low_unemploy %>%
  ggplot(aes(
    x = unemployment_rate,
    y = reorder(NAME, unemployment_rate)
  )) +
  geom_col()

low_unemp_plot
```

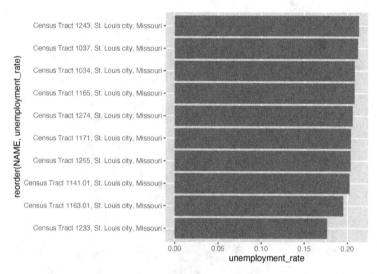

Let's clean up these labels and add a title for the two plots. We do this through the labs() function. We do not have to re-write all the code to create the plot since we stored it in a variable. All we will have to do is to add the labels layer through using + and then labs().

```
high_unemp_plot <- high_unemp_plot +
  labs(
    title, "Ten census tracts with the highest unemployment rate",
    x = "Unemployment Rate",
    y = "Census Tract"
  )

high_unemp_plot
```

We will also do the same for `low_unemp_plot`.

```
low_unemp_plot <- low_unemp_plot +
  labs(
    title = "Ten areas with the lowest unemployment rate",
    x = "Unemployment Rate",
    y = "Census Tract"
    )

low_unemp_plot
```

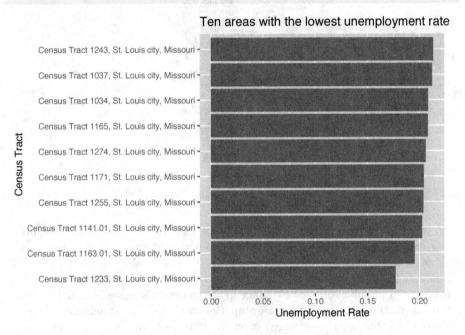

Now that we have the top ten Census block groups with the highest and lowest unemployment rates, let's see their occupations. First, let's use the `female_low_unemployment` data frame to create a plot seeing the top ten female occupations with the lowest unemployment. By getting information about these occupations, you can target your outreach by providing information and training opportunities with these occupations.

```
low_unemployment_female_plot <- female_low_unemployment %>%
  ggplot(aes(
    x = total,
    y = reorder(female_jobs, total)
  )) +
  geom_col()
```

```
low_unemployment_female_plot
```

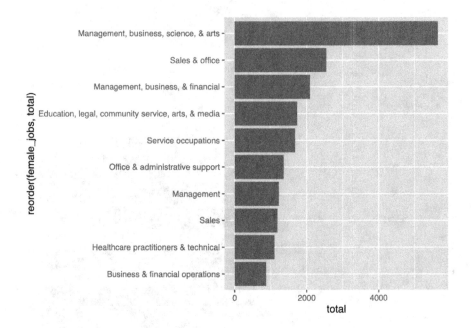

As we did in the previous set of plots, we created a variable called low_unemployment_female_plot in which we have the total number of occupations on the x-axis and the type of jobs on the y-axis. We call the reorder() function to order the bar plot by the number of jobs. The occupation with the highest number of jobs (Management, business, science and arts) is on the top of the bar plot and the occupation with the lowest number of jobs (Business & financial operations) is on the bottom of the bar plot. Let's now do the same with the male occupations.

```
low_unemployment_male_plot <- male_low_unemployment %>%
  ggplot(aes(
    x = total,
    y = reorder(male_jobs, total)
  )) +
  geom_col()

low_unemployment_male_plot
```

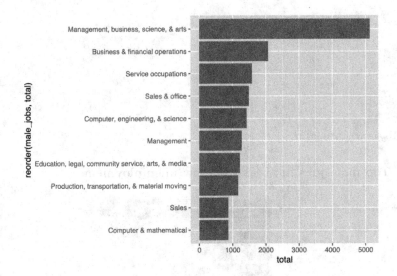

Now we will add a title and x and y axis label layer to the two plots through
+ labs().

```
low_unemployment_female_plot <- low_unemployment_female_plot +
  labs(
    title = "Top female jobs in areas with low unemployment",
    x = "Total",
    y = "Occupation"
  )

low_unemployment_female_plot
```

Top female jobs in areas with low unemployment

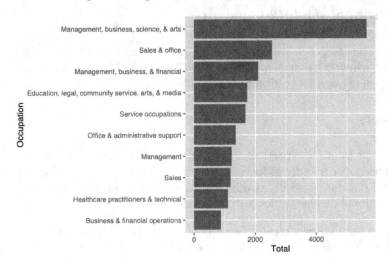

```
low_unemployment_male_plot <- low_unemployment_male_plot +
  labs(
    title = "Top male jobs in areas with low unemployment",
    x = "Total",
    y = "Occupation"
  )

low_unemployment_male_plot
```

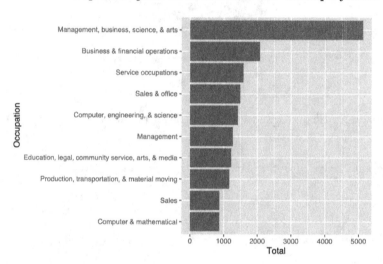

Top male jobs in areas with low unemployment

4.9 Additional *ggplot2* Concepts

This chapter aims to get you up and running by creating data visualizations using *ggplot2*, which is by no means exhaustive. The *ggplot2* library has robust features, but here are a few concepts you should know.

4.9.1 Facets

If you want to plot data by a particular categorical variable, you can use **facets**. Our data is not facet-friendly so instead, we will demonstrate facets using one of the built-in R datasets similar to our topic at hand. The gapminder dataset, a dataset on global demographics, will work with facets.

One way in which we can create facets is through the `facet_wrap()` function. Use this function when you want to facet one variable (wickham, 2016).

Let's look at the life expectancy (lifeExp) vs. GDP per capita (gdpPercap) by decade from 1957 to 2007. Before we can start faceting, we must get the data.

```
# saving the gapminder data
gapminder_data <- gapminder

# separating the gapminder data by decade
gapminder_decade_data <- gapminder_data %>%
    filter(year %in% c(1957, 1967, 1977, 1987, 1997, 2007))
```

We create a variable called `gapminder_data` by calling the built-in dataset called gapminder. We want the data by decade from 1957 to 2007, so we create another variable called `gapminder_decade_data` and we filter the data by decade from 1957 to 2007 using the `%in%` operator to extract the specific years that is indicated in the vector `c(1957, 1967, 1977, 1987,1997, 2007)`.

```
gdp_lifeexp_year_plot <- gapminder_decade_data %>%
    ggplot(aes(gdpPercap, lifeExp)) +
    geom_point() +
    facet_wrap(~year)

gdp_lifeexp_year_plot
```

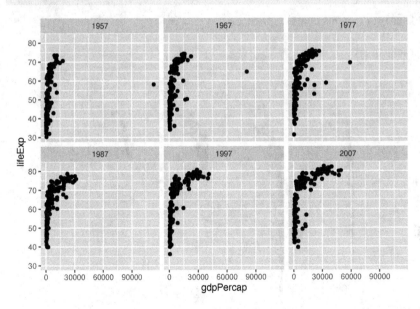

Now we will create a variable called `gdp_lifeexp_year_plot` that will store the plot with GDP per capita on the the x-axis and life expectancy on the y-axis. We will make this plot a scatterplot by adding `geom_point()` and facet by year through the `facet_wrap()` function. The year inside the `facet_wrap()` function indicates that you want to facet by year.

You can use `facet_grid()` to facet your plot using two variables (Wickham and Grolemund, 2016). Let's create a plot that shows the relationship between the GDP per capita (gdpPercap) and life expectancy (lifeExp) in the Americas and Europe by decade from 1957 to 2007. First, we need to filter the `gapminder_decade_data` to only include data from the Americans and Europe. We will then use the `facet_grid()` function to facet the plots by continents. We supply a variable before and after the ~.

```
gdp_lifeexp_dec_plot <- gapminder_decade_data %>%
    filter( continent == "Americas" | continent == "Europe") %>%
    ggplot(aes(lifeExp, gdpPercap)) +
    geom_point() +
    facet_wrap( continent ~ year)

gdp_lifeexp_dec_plot
```

To spread values across columns, we would need to use `.~` (Navarro and Pedersen, 2020). In the case of our data, to do this by continent, we would need to use `facet_grid(. ~ continent)`

```
gdp_lifeexp_plot <- gapminder_decade_data %>%
    ggplot(aes(gdpPercap, lifeExp)) +
    geom_point() +
    facet_grid(. ~ continent)

gdp_lifeexp_plot
```

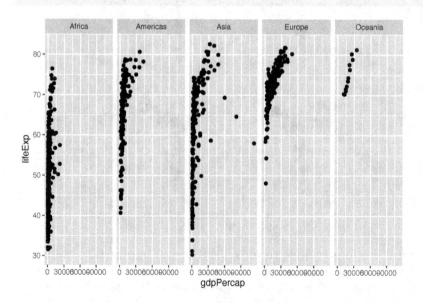

In this code chunk, we created a variable to store our plot called gdp_lifeexp_plot in which life expectancy (lifeExp) is on the y-axis and (gdpPercap) is on the x-axis. As you can see, the plot shares the same y-axis which allows for easy comparison across columns. Likewise, we can spread values across rows using ~. Let's flip the plot from above where we can spread GDP per capita and life expectancy by continent and compare row-wise.

```
gdp_lifeexp_plot <- gapminder_decade_data %>%
  ggplot(aes(gdpPercap, lifeExp)) +
  geom_point() +
  facet_grid(continent ~ .)


gdp_lifeexp_plot
```

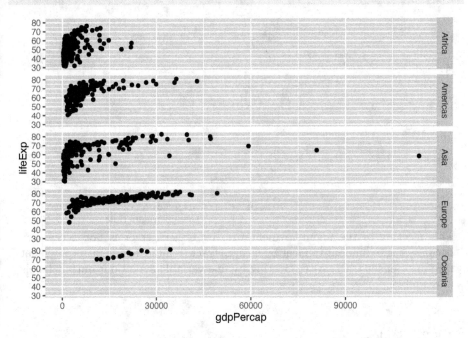

With this plot, we can see that the variables share the same x-axis that allows us to compare row-wise. However this plot doesn't seem as aesthetically pleasing as the previous one since it uses too much space row-wise. It's important to test which type of faceting is appropriate for your data.

Now that we know about facet_grid() and facet_wrap(), let's revisit why the data we have been using isn't appropriate for faceting. Our data isn't appropriate for faceting because we do not have any categorical variables in our data so we can not do any subsetting. Each variable in our data has one value attributed to it; each Census tract and occupation has one value attributed to it. To best show this, let's create a facet grid in faceting the unemployment rate by Census tract using the NAME variable through `facet_grid()`.

```
high_unemp_plot <- high_unemp_plot +
    facet_grid(~NAME)
```

```
high_unemp_plot
```

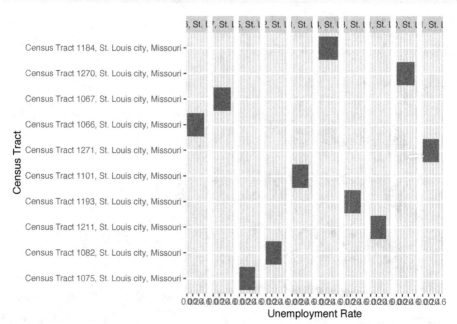

This faceting of the unemployment rate by Census tract is not neccessary because it is redundant to facet by Census tract to indicate one unemployment rate per Census tract. Also, faceting by one variable makes our output labels unreadable.

4.9.2   Scales

Scales change how the plot looks. Changing the scales, such as the minimum and maximum values in the x and y axes and the data breaks, can make your visualization more readable. For example, in the last plot of the top ten female occupations in Census tracts with the lowest unemployment, the data is broken up by 2,000 units, with the max range being 6,000. Let's change the scale in which the data is broken up by 1,000 units, with the max range being 7,000.

```
low_unemployment_female_plot <- low_unemployment_female_plot +
  scale_x_continuous(breaks = seq(1000, 7000, by = 1000)) +
  labs(
    title = "Top female jobs in areas with low unemployment",
    x = "Total",
    y = "Occupation"
  )

low_unemployment_female_plot
```

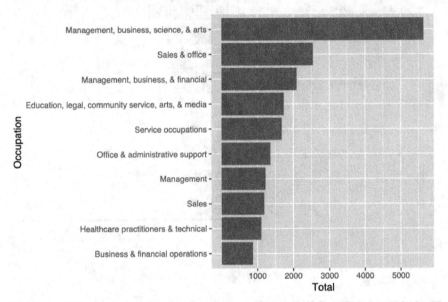

We call the low employment plot and build upon it by adding a scale layer using + and the scale_x_continous() function. within that function is the code seq(1000, 7000, by = 1000) in which the minimum x-axis value is 1,000 and the maximum value is 7,000 and each break in the x-axis by 1,000.

4.9.3 Themes

If you do not like the default **theme** of your plot, there are several built-in themes that you can use. For more themes, you can use the *ggthemes*[3] package by Jeffrey Arnold. Let's change the plot on the top ten male occupations

---

[3]https://github.com/jrnold/ggthemes

in Census tracts with the lowest unemployment to `theme_classic()`, which
removes the grid lines.

```
low_unemployment_male_plot <- low_unemployment_male_plot +
    theme_classic()
```

```
low_unemployment_male_plot
```

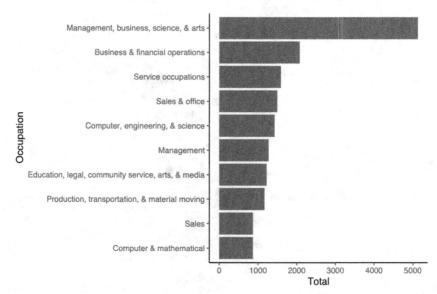

4.10   Summary

In this chapter we learned how to create data visualizations with the *ggplot2*
package. We also learned how to use the *tidycensus* package in accessing Census data within RStudio which involves accessing the Census API through a
Census API key. *ggplot2* was created based on the "Grammar of Graphics"
in which one graphic element is layered on top each other. While there are
various types of visualizations that one can create with *gglot2* we focused on
creating bar plots with `geom_col()` and scatter plots with `geom_point()`. We
were able to explore these functions through creating bar plots of Census tracts
that has the lowest rate of unemployment based on gender using *tidycensus*.
We also created scatterplots using `geom_point()` with gapminder data. With

these plots, we were able to further explore various *ggplot2* functions, such as facets, scales, and themes.

4.11   Additional Resources

- Data visualization with *ggplot2* cheatsheet: https://posit.co/resources/cheatsheets/
- *ggplot2: Elegant Graphics for Data Analysis*: https://ggplot2-book.org/index.html

---

4.12   Further Practice

- For the high and low unemployment plots, change the plot's theme.
- Explore the built-in R `datasets` through the `datasets` package. Choose a dataset and create facets of a particular variable.

5

Webscraping with rvest

5.1 Learning Objectives

1. Explain how to determine which HTML tags will provide the needed data
2. Use the *rvest* package to scrape webpage content into the RStudio IDE
3. Create a dataset from the scraped content
4. Use *dplyr* to normalize the dataset
5. Create a CSV file from the scraped content

5.2 Terms You'll Learn

- web scraping
- subset

5.3 Scenario

You want to advertise your report and outreach program to garner support from government leaders in St. Louis. Reaching out to the aldermen's offices within the city seems most likely to support this goal. A website lists the aldermen, but you have to click on each name to get to the webpage that lists each alderman's email address. Rather than clicking on 29 links and copy/pasting each email address, name & ward number, you can use *rvest* to scrape this information from the city's website and put it into a table you can update and reuse through the life of this program.

DOI: 10.1201/9781003218012-5

5.4 Packages & Datasets Needed

First, we need to load the packages we need into our current environment to access the functions. We are using *rvest* and *purrr* in this chapter while revisiting *dplyr*.

```
library(tidyverse)
library(rvest)
library(xml2)
```

5.5 Introduction

So much valuable information and data are available on the internet, but it's hard to use directly on a webpage. Getting information from a website in an automated manner and into a format that's usable and compatible with other applications is made much easier with the advent of web scraping. **Web scraping**[1] uses code to load a webpage and extract data contained on the page. Taking a list of names and addresses and putting it into a table that you could turn into a set of mailing labels without repetitive copy/paste keystrokes is one way that web scraping can automate an onerous task. Contact information for a mailing list is a straightforward use case for web scraping, but the options are as numerous as web pages on the internet.

Web scraping works by selecting HTML or CSS fields on a webpage and using code to extract the values of those fields. Often, a web page will use a /table HTML tag to indicate tabular data, but just as often, websites use CSS to make content appear in a table without utilizing the semantic HTML option of the table tag; it looks great, but it can be hard to scrape. As a user, it's up to you to figure out what tags to scrape to get the data you need. Determining which tags to scrape can be challenging depending on the website's code, but it also allows you to choose only those elements that matter. The proper tags with a bit of code can return reams of data in seconds that could take hours to extract manually. The ability to pull content from websites significantly broadens what users can think of as data.

---

[1]https://en.wikipedia.org/wiki/Web_scraping

5.6   Identifying & Scraping Website Components

To get in contact with each alderman, we need their email. But to keep our records straight, we also want to know the alderman's name and which ward they represent. Thus, the metadata fields we need are ward number, alderman name, and alderman email address.

The website listing St. Louis aldermen is https://www.stlouis-mo.gov/government/departments/aldermen/Wards-1-28.cfm. It lists the 28 city wards and the alderman's name representing that ward.

Wards and Aldermen

Links to Wards 1 through 28, and their Respective Aldermen

New Ward Boundaries and Representation

Representation based on new ward boundaries will begin after the aldermanic general election on April 4, 2023. Learn more about redistricting.

All Wards

Each ward and its associated Alderman is listed below. Click on a ward or Alderman for details.

Learn more about the Board of Aldermen.

Wards and Aldermen

Ward	Alderman
Ward 01	Sharon Tyus
Ward 02	Lisa Middlebrook
Ward 03	Brandon Bosley
Ward 04	Dwinderlin Evans
Ward 05	James Page
Ward 06	Christine Ingrassia
Ward 07	Jack Coatar
Ward 08	Annie Rice
Ward 09	Dan Guenther
Ward 10	Joseph Vollmer

Find Your Aldermen

Street Address or Parcel:

Search

Ward Maps

FIGURE 5.1 List of St. Louis wards & aldermen.

The Wards & Aldermen page provides us with the ward numbers and their representative alderman, but it does not list the alderman's email address. Both the ward number and the alderman name are hyperlinks to the same alderman's webpage.

The individual alderman websites do provide the alderman's email address, physical address, and phone number. From these pages, we can anticipate two

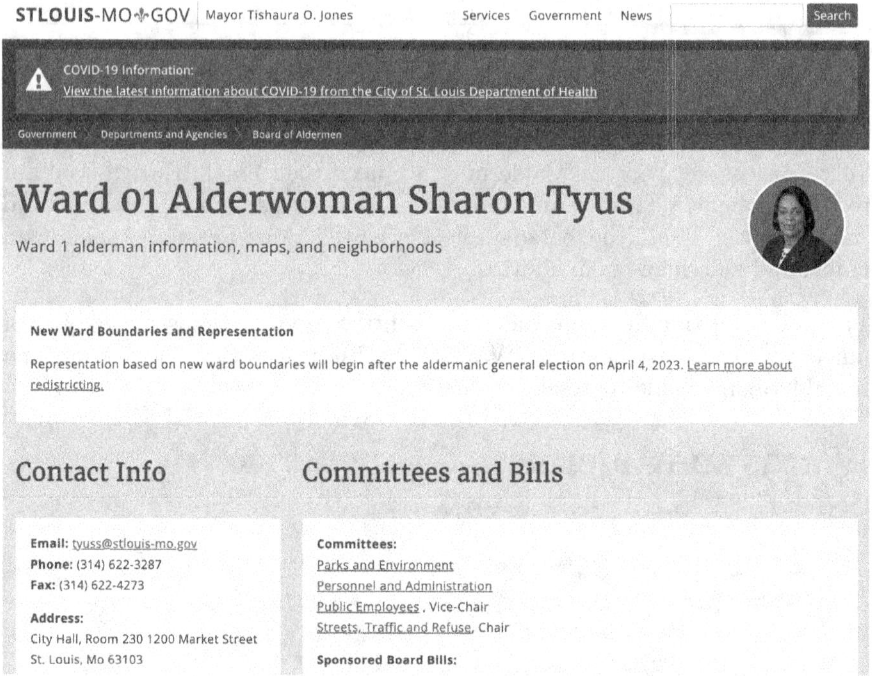

FIGURE 5.2 Ward 1 Alderman Sharon Tyus' main webpage.

stages to our web scraping. First, we'll need to scrape the list of wards and aldermen. Then we'll need to scrape each alderman's webpage for their email address.

Before we get started coding, we need to determine the CSS tags, or selectors, that we need to scrape. To do that, we'll utilize a tool called Selector Gadget[2], which is available as a Chrome[3] browser extension, to expose and identify CSS tags on the data we need. The *rvest* website[4] recommends SelectorGadget.

After installing Chrome and SelectorGadget, we navigate to https://www.stlouis-mo.gov/government/departments/aldermen/Wards-1-28.cfm and click SelectorGadget from the Chrome extensions menu. Once SelectorGadget is engaged, we click on the content we want to scrape (any ward number or alderman name). SelectorGadget highlights the content and provides the CSS selector to scrape in the dialog box at the bottom of the browser. If Selector-Gadget highlights more fields than what we want, we'll click on the extraneous elements to deselect them. Once finished, SelectorGadget will display the se-

---

[2]https://selectorgadget.com/
[3]https://www.google.com/chrome/
[4]https://rvest.tidyverse.org

lector we need to use with *rvest*. We'll need to repeat this for each page and element we need.

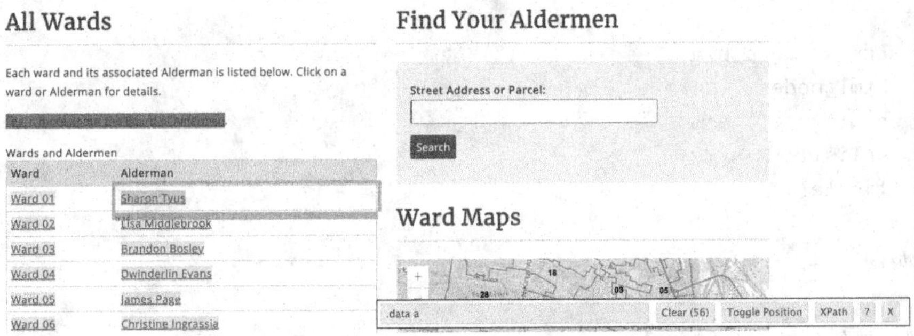

FIGURE 5.3 SelectorGadget on the Wards & Alderman webpage.

To scrape the aldermen's names and ward numbers, we need the CSS selector `.data a`.

5.7 Web Scraping Part 1: Wards & Aldermen

1. Scrape each alderman's name and ward number

First, we will use the `read_html()` function to read the webpage we want to scrape. Our second step is to save that scraped webpage in an HTML file so that we can access it later.

```
scrape <- read_html(
  "https://www.stlouis-mo.gov/government/departments/aldermen/
    Wards-1-28.cfm"
)

write_html(scrape, "scrape.html")
```

After we read the scraped HTML file back in, then we'll use `html_nodes()` to indicate which selector to scrape (`.data a`). The `html_text()` function turns the scraped content into readable text format. We'll `unlist()` the data and put it into an R table format with `tibble()`. In Chapter 2 we talked about different data structures in R; `base::unlist()` is an R function that converts a list into a vector. This standardizes the data, ensuring that all elements have

the same data class so that they can be moved into a table with the `tibble()`
function. Finally, we can check our dataset to make sure we scraped what we
wanted:

```
wards <- read_html("scrape.html") %>%
  html_nodes(".data a") %>%
  html_text() %>%
  unlist() %>%
  tibble()

wards
```

```
## # A tibble: 58 x 1
##      .
##      <chr>
##   1  "\nWard\n01\n"
##   2  "\nSharon Tyus\n"
##   3  "\nWard\n02\n"
##   4  "\nLisa Middlebrook\n"
##   5  "\nWard\n03\n"
##   6  "\nBrandon Bosley\n"
##   7  "\nWard\n04\n"
##   8  "\nDwinderlin Evans\n"
##   9  "\nWard\n05\n"
## 10  "\nJames Page\n"
## # ... with 48 more rows
```

2. Tidy aldermen names and ward numbers

We can see from the output above that the `wards` object doesn't have the ward
number and alderman name on a single line, as we need it to be. The problem
is that because the ward number and the alderman's name use the same CSS
selector, R put all columns and rows into the same character vector without
unique separators between lines. To fix this problem, we need to create a new
tibble where column 1 is all the odd rows of wards and column 2 is all the
even rows, in order to have a tidy dataset[5].

We'll need to use two new operators to select the even and odd rows separately:
`%%` and `==`. The double equal signs mean that two values must be exactly equal
to each other, and the double percentage signs return the remainder (as in
division). We're creating objects called even when rows are divisible by 2 (the
remainder is zero) and odd when a row has a remainder of 1 when divided

---

[5]https://vita.had.co.nz/papers/tidy-data.pdf

by 2. We will do this in two steps to create an object for our odd rows and another for even rows:

```
ward_num <- wards %>%
  filter(row_number() %% 2 == 1) %>% # odd rows
  tibble() %>%
  rename(Ward = names(.))

# the '.' can be read as "the current object,
# created in the previous line"
```

Additionally, we'll rename the columns with a blank header using the dplyr rename() function, because columns without header rows are hard to work with and hard to remember what's inside that column. When using rename(), we need to pass the function two pieces of information: the data and the new column name. However, since we are using the pipe to combine functions, the data isn't an object we can reference so we need a workaround. The argument we'll pass to rename() here is the new name and we'll say that it is equal to the name that is currently called . using the base::names() function, which R recognizes applies to the name of a list or vector.

While this explanation is a little long, this code is more efficient than creating an object with an blank column name and then adding a new object that is the same data, but with a new column name. In essence, this code allows us to do all of our actions in one object rather than two.

```
alderman <- wards %>%
  filter(row_number() %% 2 == 0) %>% # even rows
  tibble() %>%
  rename(Alderman = names(.))
# rename the Alderman column, which has a blank column name
```

Finally, the column bind function, base::cbind(), will let us bind the objects for alderman names and ward numbers together into one table, where each vector becomes a column in the table:

```
wards_alderman <- cbind(ward_num, alderman)
# combine the two columns together into one tibble

head(wards_alderman)
```

```
##            Ward              Alderman
## 1 \nWard\n01\n          \nSharon Tyus\n
## 2 \nWard\n02\n      \nLisa Middlebrook\n
```

```
## 3 \nWard\n03\n        \nBrandon Bosley\n
## 4 \nWard\n04\n        \nDwinderlin Evans\n
## 5 \nWard\n05\n               \nJames Page\n
## 6 \nWard\n06\n \nChristine Ingrassia\n
```

We now have a tidy dataset where each observation includes the ward number
and that ward's alderman.

5.8   Web Scraping Part 2: Email Addresses on Individual Pages

1. Scrape the Aldermans' contact page URLs

The first page we scraped lacks the email address for each alderman, but we
did scrape the name and ward number. Each alderman's web page contains
their contact information, including their email address. However, we need to
scrape the links to each alderman's page before scraping the email addresses.
With the second round of scraping, we will pull the URLs (href) for the
aldermen's pages and each alderman's email address from that page.

We're still interested in the same CSS selector, .data a, but this time
we do not want the value for that selector. Using html_attr(), we indicate that we want the link, or href, for those values. The href doesn't include the entire URL to the aldermen's sites but only the relative file path:
/government/departments/aldermen/ward-1. Each file path is relative to the
base_url in that the relative file path is text that would be appended to the
base_url. Thus, our code will need to scrape the href but then combine that
with the base URL, https://www.stlouis-mo.gov/, to find the full URL we
need to scrape in the subsequent step. The R function file.path() is a faster
version of paste, and we will use it to paste the two URL components together. In coding, the relative file path is usually represented by ., as we see
in the file.path(base_url, .). When run repeatedly, the relative file path
will change, but each will be combined with the base_url:

```
base_url <- "https://www.stlouis-mo.gov"

href <- read_html("scrape.html") %>%
```

```
html_nodes(".data a") %>%
html_attr("href") %>%
file.path(base_url, .) %>%
str_replace_all("gov//", "gov/") %>%
tibble()
```

There's an additional step to mention here: the base URL ends in a forward slash, and the relative file path starts with a forward slash. Combining these two portions will get a double forward slash that will invalidate the complete URL. We'll utilize a function from the *stringr* package (part of *tidyverse*) that lets us modify a string in many ways. We want to use `str_replace_all()` to remove the double forward slash. However, we must specify that we only want to replace the double forward slash after gov. Otherwise, our code will replace the double forward slash after `https:`. Only then can we store the data in a tibble and check to see that the data looks clean.

```
head(href)
```

```
## # A tibble: 6 x 1
##    .
##    <chr>
## 1 https://www.stlouis-mo.gov/government/departments/al~
## 2 https://www.stlouis-mo.gov/government/departments/al~
## 3 https://www.stlouis-mo.gov/government/departments/al~
## 4 https://www.stlouis-mo.gov/government/departments/al~
## 5 https://www.stlouis-mo.gov/government/departments/al~
## 6 https://www.stlouis-mo.gov/government/departments/al~
```

2. Clean up the website addresses

Similar to the duplication problem we had at the start of the chapter, we're getting duplicate lines for each URL because both the ward number and the alderman's name link to the same contact page on the original web page. Both also use the same CSS tag, so we can't separate the two and only pull one of the links. We need to pull out only one copy of each link using the previous filtering by even rows. Additionally, as with so many of our columns, we need to rename those with empty header rows (which include only a period):

```
urls <- href %>%
  filter(row_number() %% 2 == 0)
```

```
websites <- rename(urls, website = .)

websites
```

```
## # A tibble: 29 x 1
##      website
##      <chr>
##   1 https://www.stlouis-mo.gov/government/departments/a~
##   2 https://www.stlouis-mo.gov/government/departments/a~
##   3 https://www.stlouis-mo.gov/government/departments/a~
##   4 https://www.stlouis-mo.gov/government/departments/a~
##   5 https://www.stlouis-mo.gov/government/departments/a~
##   6 https://www.stlouis-mo.gov/government/departments/a~
##   7 https://www.stlouis-mo.gov/government/departments/a~
##   8 https://www.stlouis-mo.gov/government/departments/a~
##   9 https://www.stlouis-mo.gov/government/departments/a~
## 10 https://www.stlouis-mo.gov/government/departments/a~
## # ... with 19 more rows
```

3. Combine the ward numbers, alderman names, and website URLs

We've come far enough that it makes sense to combine the objects we have into a single table. Again, we'll use the column bind function to combine the objects we have so far into one table:

```
all_urls <- cbind(wards_alderman, websites)
# one table of ward, aldermen, and links to their webpages
```

5.8.1 Automating Tedious Scraping

As is often the case with data "in the wild," the information we need about the city aldermen isn't neatly shown on one webpage. We scraped the alderman names, ward numbers, and personal website URLs, but clicking on 28 websites and copying and pasting email addresses sounds tedious and prone to error. Plus, aldermen join and leave this list, making our data valid for only a limited time.

Thankfully, just as we can automate data collection by scraping it off a website, the Tidyverse presents a package and function combination to automate tedious web scraping tasks. In this case, we'll use the Tidyverse package *purrr* to iterate over the list of website URLs. The map() function from *purrr* lets us perform the same commands we did previously, by iterating on each alderman's website. For every website listed in `vector`, we will apply `read_html()` to read in those websites. Then we will scrape the email HTML field from each page using `html_node()`.

However, there is one problem with our data as it currently stands. R has several object classes and `purrr::map()` only works on lists or atomic vectors. Other object classes include matrices, arrays, factors, and data frames. Having an object of the wrong type for the function you want to execute is a common cause of error messages when coding in R. To find out the object class you have, you can use `base::class()`:

```
class(all_urls)
```

```
## [1] "data.frame"
```

Our object is a data frame, and while that is great for creating tables, it will not work for `map()`. We only need to change the column we need to use in our scraping task, so we will select the third column, `websites`, from the object `all_urls`. To do that we will **subset**, or take one group of data within a larger group, the third column of `all_urls`. The double square brackets on either side of the number three indicate what is being subsetted. This is a useful operation that we'll use several times in this book when we need to break out one part of our larger dataset. We'll create a new object, `vector_urls`, where we change the data structure of the `websites` column from a data frame to a vector using `as.vector()` so that the scraping function we need to do next will work.

```
vector_urls <- as.vector(all_urls[[3]])
```

1. Scrape aldermen email addresses

Before scraping email addresses for each alderman, we must determine which CSS selector we need to scrape. Using SelectorGadget as we did before, we see that each alderman's email address uses the `strong+` a selector.

We are now ready to automate scraping each alderman's website for their email address. First, we will use `map()` to read in the HTML for each element of our vector of alderman website URLs. Then we will use `map()` again to scrape the CSS selector `strong+` a from the HTML we read for each site.

```
scrape_sites <- map(vector_urls, read_html)
# scrape each URL in the vector
```

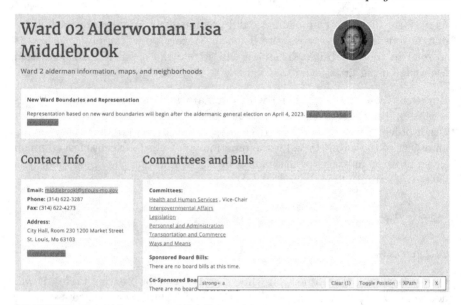

FIGURE 5.4 Alderwoman Middlebrook's contact page.

```
scrape_email <- map(scrape_sites, html_node, "strong+ a")
# pull out the email address on each of those pages
```

    2.   Convert email addresses into readable format

We will use map() again to turn each scraped field into a text format using html_text() and then put the clean text into a table.

```
email_list <- map(scrape_email, html_text) %>%
  unlist() %>%
  tibble()
```

5.9   Create Reusable Data File

To reuse the scraped data in subsequent chapters, we need to make a new table with clear column headers that combine all four tables into one: ward

number, alderman name, and alderman website from `all_urls`, and alderman
email.

1. Rename column headers

```
emails <- rename(email_list, Email = .)
```

2. Create a single table with all necessary elements

```
complete <- cbind(all_urls, emails)
head(complete)
```

Now have the full dataset we need to contact all the aldermen!

3. Write to CSV

Once we've created a complete, tidy table of all the wards, the aldermen, and
their email addresses, we want to save that as a CSV file for reuse or sharing.

```
write_csv(complete, "data/aldermen-contact.csv")
```

The `write_csv` function takes our object, "complete," and turns it into a CSV
file we could read back into R or view in a spreadsheet application like Google
Sheets or Microsoft Excel.

5.10 Summary

We created a CSV file of aldermen's contact information by identifying which
webpages had the information we needed, what specific HTML tags that data
was stored in, scraping those particular tags, and tidying the scraped data.
The St. Louis aldermen webpages are a real-life example of dispersed data
on the web that can't be copied and pasted into a spreadsheet. Instead, we
used code to speed up and automate data acquisition, saving the code and
the output. When new aldermen are elected, we need only re-run the code to
update the contact list, which saves us a lot of time and energy.

5.11   Further Practice

- scrape each alderman's phone number and add it to the `complete` object as a new column
- scrape a webpage from your employer that lists employees
- scrape the contact information for your state senators

5.12   Additional Resources

- Apply functions with *purrr* cheatsheet: https://posit.co/resources/cheatsheets/
- *rvest* package site: https://rvest.tidyverse.org/
- *Webscraping with R* by Steve Pittard: https://steviep42.github.io/webscraping/book/

6

Mapping with tmap

6.1 Learning Objectives

1. Define geographic concepts relevant to cartography such as geographic coordinate system, projected coordinate system, datum, and spheroid.
2. Describe vector and raster data.
3. Perform loading spatial and tabular data into RStudio using relevant functions.
4. Apply a data transformation of spatial data into an appropriate projection.
5. Use appropriate functions in the tmap package to display spatial data.

6.2 Terms You'll Learn

- Spatial data
- Vector data
- Raster data
- Projection
- Latitude
- Longitude
- Shapefile
- Census tract

DOI: 10.1201/9781003218012-6

6.3   Scenario

After creating plots of occupations with the lowest unemployment, you decide
to create maps showing the total unemployment in the city. Being able to
see the areas with the lowest and highest unemployment will be helpful in
targeting outreach in areas with the highest unemployment. You want these
maps to be interactive so users can click on an area and get information about
that area in the city. However, you are not sure how to visualize this in a map.
Questions that come to mind are

- Where can I get spatial data to make a map?
- What kind of spatial data do I need to make a map?
- What relevant R packages can be used to create these maps?

6.4   Packages & Datasets Needed

```
library(tidyverse)
library(tidycensus)
library(tmap)
library(sf)
library(raster)
```

6.5   An Overview of Spatial Data

To make a map, you need to use **spatial data**. So what makes spatial data
special? Spatial data has geographic attributes which allow it to be mapped.

6.5.1   Vector and Raster Data

There are two types of spatial data: vector data and raster data. **Vector data**
is composed of lines, polygons, and points. Lines can be used to represent

features such as roads, county boundaries for polygons, and the location of
bus and metro train stops for points data.

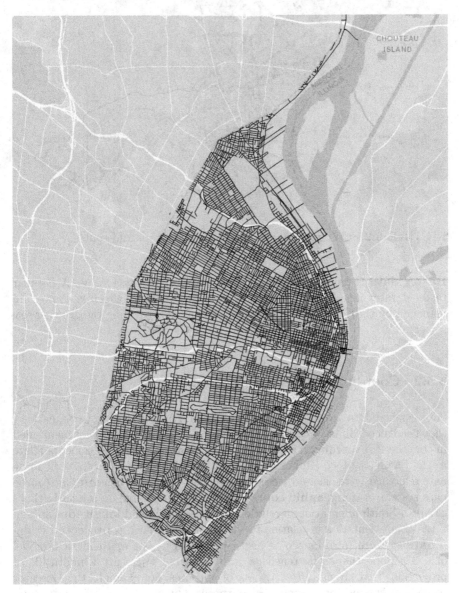

FIGURE 6.1 St. Louis city Streets.

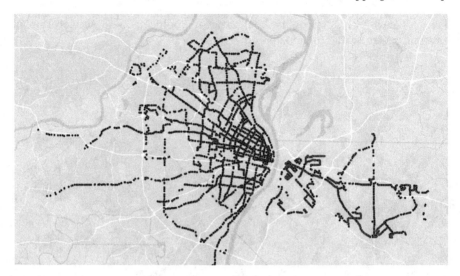

FIGURE 6.2 St. Louis MetroBus & MetroLink stops and routes.

Raster data is image data, where each pixel represents a value. A popular raster dataset is the National Aerial Imagery Program[1] which is a program managed by the USDA that collects satellite imagery during the growing seasons.

6.5.2   Coordinate Reference Systems

Some spatial data are not ready to be mapped. For spatial data to be mapped, they need a coordinate reference system, or coordinate system, first mentioned in chapter 4. A coordinate reference system shows how spatial elements relate to the Earth's surface (Robin Lovelace, 2019). There are two types of coordinate reference systems: a projected coordinate system and a geographic coordinate system. A **geographic coordinate system** is based on **latitude** (lines of North-South orientation in relation to the equator) and **longitude** (lines of East-West orientation in relation to the Prime Meridian). A **projected coordinate system** (also known as just a **projection**) is a mathematical model of a 3D globe that is flattened on a 2D surface. There are a multitude of projections that are used for a variety of purposes. For example, the Mercator projection is good for navigational purposes, but not so good for visualizing the Earth on a 2D surface since it exaggerates size of continents such as North America and minimizes the size of continents such as Africa. In the context of

---

[1] https://www.usgs.gov/centers/eros/science/usgs-eros-archive-aerial-photography-national-agriculture-imagery-program-naip

FIGURE 6.3 St. Louis city parks.

FIGURE 6.4 NAIP Imagery of St. Louis.

our scenario, it is best to use the State Plane Projection[2] since this projection
is highly accurate for the local level.

6.5.3   Is Your Data Projected or Not?

How do you know if your data is projected? One way is to simply look at a
map. For example, an unprojected map of the US would have straight lines
and the states would look distorted. For example, the northern part of the
country would be mostly a straight line. A projected map of the US reflects
the curvature of the earth and the northern border of the country would be
curved.

Another way that you know that data isn't projected is that you will get errors
trying to run specific spatial analyses. For example, if you wanted to calculate
the nearest bus stop from the library and you are using unprojected data, you
will get an error message. Since the data is not projected, there is no spatial
reference, which means that you cannot do any spatial calculations.

---

[2]https://www.usgs.gov/faqs/what-state-plane-coordinate-system-can-gps-provide-
coordinates-these-values

In R, you can use the `st_crs()` function in the *sf* package to determine whether your spatial data has a projection or not. Let's see if the US shapefile below has a projection:

```
us <- st_read("us_state_clip/us.shp") %>%
  st_crs()
```

```
## Reading layer `us' from data source
##   `/Users/sarahlin/r4lis-crc/us_state_clip/us.shp'
##   using driver `ESRI Shapefile'
## Simple feature collection with 49 features and 14 fields
## Geometry type: MULTIPOLYGON
## Dimension:     XY
## Bounding box:  xmin: -124.8 ymin: 24.4 xmax: -66.89 ymax: 49.38
## Geodetic CRS:  NAD83
```

This shapefile does not have a projection. Let's see what another shapefile looks like.

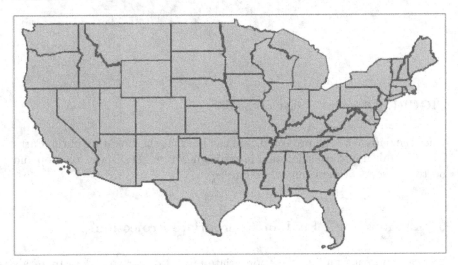

FIGURE 6.5 Unprojected map of the US.

You can tell that this shapefile does not have a projection given the straight lines at the top and bottom of the country. On a spherical surface, the US would not look like this. This is where projections come into play.

To change the projection, we will have to use `st_transform()` and we will use the US National Atlas Equal Area projection[3].

---

[3]https://epsg.io/2163

```
us_project <- us %>% st_transform(crs = 2163)
tmap_mode("plot")
tm_shape(us_project) +
  tm_polygons()
```

FIGURE 6.6 Projected map of the US.

Once the data has been projected, you can see that the curvature of the Earth has been taken into consideration in projecting it on 2D surface. The top and bottom borders of the country are curved.

6.5.4 How Can I Find an Appropriate Projection?

You can approach finding the appropriate projection for your data by establishing the purpose of using your spatial data. Are you using spatial data for analysis purposes or to display something? After you determine that, you should think about your area of interest. On what scale are you doing your mapping and analysis? In terms of trying to find the right projection for your area of interest, you can use the epsg.io[4] website to determine your projection based on location. After that you can create a list of possible projections and narrowing it down based on your purpose and further research (see additional resources at the end of this chapter).

---

[4]http://epsg.io

In the context of this project, we will be using the NAD 1983, State Plane Missouri East, FIPS 2401 feet. As mentioned before, the state plane projection is good in mapping local areas given the way the zones are derived.

In short, to map data you need: 1. A spatial dataset. 2. An appropriate projection for that spatial dataset.

This overview of spatial data is just to get you up and running. For more in-depth information about coordinate reference systems and other spatial data foundational concepts, please refer to the "Geographic Data in R"[5] chapter in *Geocomputation in R* by Robin Lovelace, Jakub Nowosad, and Jannes Muenchow (Robin Lovelace, 2019).

6.6 Loading in the Data

The type of spatial data that we will be working with is called a **shapefile**. A shapefile is a vector data format that holds the spatial object, and the attribute information of that spatial object. It is actually made of up several files such as a database file, an xml file that holds the metadata, and a projection file (if the shapefile already has a projection). We are going to load in the WARDS_2010 shapefile which was retrieved from the St. Louis City open data portal[6]. If you take a look at the WARDS_2010 shapefile, you will see that it is actually comprised of several files such as WARDS_2010.prj, which holds the projection, and WARDS_2010.dbf that holds the attribute information for each ward.

We will need to use st_read() function from the *sf* package to load the shapefile into RStudio. After that, we will use the st_crs() function to see if the shapefile has a coordinate system. We are also going to read the unemployment_tract shapefile that was created in the previous chapter.

```
stl_wards <- st_read("nbrhds_wards/WARDS_2010.shp")
stl_tracts <- st_read(
  "unemployment_tract/unemployment_tract.shp"
) %>%
  rename("unemployment_rate" = "unmply_") %>%
  mutate(unemployment_rate = unemployment_rate * 100)
```

---

[5]https://geocompr.robinlovelace.net/spatial-class.html#crs-intro
[6]https://www.stlouis-mo.gov/data/

We can see that the shapefile has a coordinate system which is NAD 1983, State Plane Missouri East, FIPS 2401 feet. Next we will create a simple static map using the *tmap* package. We will call the stl_wards shapefile within the tm_shape() function which is used to call a spatial object. Within that function we will designate miles as our units of measurement. The stl_wards shapefile is a polygon, so we need to also add the tm_polygons() function to draw the polygons.

```
tmap_mode("plot")
tm_shape(stl_wards, unit = "mi") +
  tm_polygons()
```

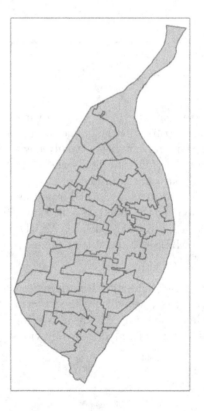

FIGURE 6.7 St. Louis city wards.

We can add cartographic elements to the map which includes a title, scale bar, and north arrow using the tm_layout(), tm_scale_bar(), and tm_compass() functions, respectively.

```
tmap_mode("plot")
tm_shape(stl_wards, unit = "mi") +
  tm_polygons() +
  tm_layout(title = "Wards in St. Louis City", title.size = .7) +
  tm_scale_bar(position = c("right", "bottom"), width = .3) +
  tm_compass(position = c("left", "top"))
```

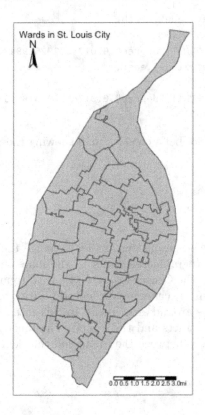

FIGURE 6.8 Wards in St. Louis City with cartographic elements.

Let's make a static map of the Census tract map. A **Census tract** is one of the smallest statistical units of a county and has about an average population of 4,000 (U.S. Census Bureau, 2022) First, let's check and see what the coordinate system of the Census tracts.

```
st_crs(stl_tracts)
```

```
## Coordinate Reference System:
##   User input: NAD83
```

```
##    wkt:
## GEOGCRS["NAD83",
##      DATUM["North American Datum 1983",
##          ELLIPSOID["GRS 1980",6378137,298.257222101,
##              LENGTHUNIT["metre",1]]],
##      PRIMEM["Greenwich",0,
##          ANGLEUNIT["degree",0.0174532925199433]],
##      CS[ellipsoidal,2],
##          AXIS["latitude",north,
##              ORDER[1],
##              ANGLEUNIT["degree",0.0174532925199433]],
##          AXIS["longitude",east,
##              ORDER[2],
##              ANGLEUNIT["degree",0.0174532925199433]],
##      ID["EPSG",4269]]
```

The CRS is NAD 1983. Let's create a map showing the unemployment rate.

```
tmap_mode("plot")
tm_shape(stl_tracts) +
  tm_polygons("unemployment_rate")
```

The legend title can be cleaned up, so let's change the name of the legend
when adding the cartographic elements to the map. In addition,the legend
items are overlapping the Census tracts, so we need to make the legend items
smaller. We can do this by changing the legend position and the inner margins.
A margin is the space around an element. In this case, the element would be
the St. Louis Census tracts and the legend. The inner margins is the space
inside the frame that is between the frame and the element.

```
tmap_mode("plot")
tm_shape(stl_tracts, unit = "mi") +
  tm_polygons(
    "unemployment_rate",
    title = "Unemployment rate"
  ) +
  tm_layout(
    title = "Percentage unemployed by Census tract",
    title.size = .7,
    legend.width = 1,
    legend.text.size = .4,
    legend.title.size = .5,
    legend.position = c("left", "top"),
    inner.margins = c(0.01, 0.01, .12, .25)
```

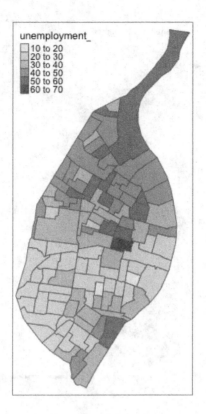

FIGURE 6.9 Percentage unemployed by Census tract map with legend.

```
) +
tm_scale_bar(position = c("right", "bottom"), width = .3) +
tm_compass(position = c("right", "center"))
```

We can make our maps dynamic and interactive with the *leaflet* package[7]. We can access that package within the tmap package itself, since tmap imports the leaflet package already. Let's create *leaflet* maps of the wards and the Census tracts. We will do that by changing the tmap_mode() function to "view." Here is a map of the wards in view mode:

```
tmap_mode("view")
tm_shape(stl_wards, unit = "mi") +
  tm_polygons()
```

---

[7]https://rstudio.github.io/leaflet/

FIGURE 6.10 Percentage unemployed by Census tract map with title and cartographic elements.

Let's create a *leaflet* version of the map showing the unemployment rates by Census tract that we created earlier. We will use the `tm_fill()` function to add color ramp to the Census tracts to indicate the unemployment rate. With *leaflet*, you can click on the Census tracts and a pop-up will appear. We can customize the pop-ups using the `popup.vars` option within `tm_fill`. The pop-up will show the unemployment rate and the name of the Census tract.

```
tmap_mode("view")
tm_shape(stl_tracts) +
  tm_fill("unemployment_rate",
    title = "Umemployment rate",
    popup.vars = c("% unemployed" = "unemployment_rate"),
    id = "NAME"
  )
```

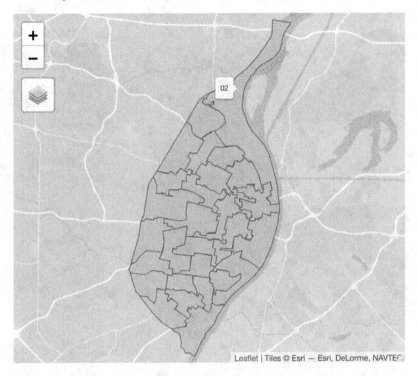

FIGURE 6.11 A *leaflet* Census tract web map.

6.7 Summary

Spatial data is composed of vector and raster data. Vector data is composed of points, lines, and polygons while raster data is image data in which each pixel represents a value. You can map vector data with the *tmap* package. One of the kinds of vector data that you can map is called a shapefile, which is actually comprises several files. In this chapter, we mapped a static map of St. Louis wards and the unemployment rate in each ward. With *tmap*, you can also create dynamic web maps using the *leaflet* package which you can call by itself or from within *tmap*. We also created a web map of the wards and the unemployment rate by ward.

FIGURE 6.12 A *leaflet* Census tract web map showing unemployment rate by ward.

6.8   Further Practice

- Modify the pop-ups to include the total population per Census tract (ttl_ppE). You can name the pop-up label "population."

6.9   Additional Resources

- U.S. Census Bureau (2022). Glossary. Retrieved from https://www.census.g ov/programs-surveys/geography/about/glossary.html#par_textimage_13
- *Geocomputation with R*: https://geocompr.robinlovelace.net/index.html

7

Textual Analysis with tidytext

7.1 Learning Objectives

1. Describe a use case for text mining
2. Use the *tidytext* package to load a textual dataset into the IDE
3. Tokenize a textual dataset
4. Perform a sentiment analysis of the textual dataset
5. Identify the text's most common words using tf-idf functions

7.2 Terms You'll Learn

- text mining
- sentiment analysis
- JSON
- for-loop
- term frequency
- inverse document frequency
- tf-idf

7.3 Scenario

Looking for a job is stressful! Being unemployed when you want to be employed is hard. Is there any data to support that unemployment has negative connotations? You'd like to add a section to your report to highlight the negative cultural connotations associated with unemployment, which job seekers feel, to further support your position that a library/community partnership is needed. In this case, newspaper articles can function as cultural data to support your premise. You'll use the *tidytext* package to text mine article metadata from the *New York Times* and add that analysis to your report.

7.4   Packages & Datasets Needed

```
library(tidyverse)
library(jsonlite)
library(tidytext)
```

7.5   Introduction

When working with datasets containing numerical values, we can use statistical analysis and visualizations to describe our data. Within librarianship, like in the humanities and social sciences, datasets are often composed of text. **Text mining**[1] provides insight into textual data by turning text, documents, or even books into a dataset. We can describe, analyze, modify, and visualize a dataset of words just as we would numbers.

We might want to analyze many academic and social texts, from our library collections, or words used in social and cultural domains that affect library patrons and the communities we serve, including some that are just for fun. Looking at a few R-specific packages for textual analysis tasks is one way to see the breadth of possibilities for text mining:

- *twitteR*[2]
- *gutenbergR*[3]
- *janeaustenR*[4]
- *scotus*[5]
- *pubmedR*[6]

---

[1] https://en.wikipedia.org/wiki/Text_mining
[2] https://cran.r-project.org/package=twitteR
[3] https://cran.r-project.org/package=gutenbergr
[4] https://cran.r-project.org/package=janeaustenr
[5] https://github.com/EmilHvitfeldt/scotus
[6] https://cran.r-project.org/package=pubmedR

- *jstor*[7]
- *nasadata*[8]
- *rfacebookstat*[9]
- *shrute*[10]

This chapter will only scratch the surface of what is possible with text mining. The first topic we'll discuss is **sentiment analysis**[11], which measures the emotional tone of a dataset, such as positive or negative. We will also touch on term frequency and measures of meaningful words. Most of the predictive search functions that librarians interact with within search engines and library catalogs use a combination of text mining and machine learning. By quantifying the relationships between words through text mining, search engine developers can use machine learning techniques to predict what will follow a term entered into a search box.

7.6 Query the NYT Article Database

To get started with text mining, we need to find a body of text that we can analyze. A useful dataset is one large enough to represent society so that sentiment analysis is meaningful; it must also be accessible. *The New York Times'* (NYT) article database fits these criteria. We can retrieve articles of particular interest for mining with the NYT developer API. We first discussed APIs in Chapter 3. Accessing an application, like an article database, via code is really useful when you need to pull a large amount of data or make repeated queries from a database.

Accessing the article database through its API is not the only way to access that data. There is an R package specific to the NYT articles database called *nytimes*. However, its last update was several years ago. We also found it lacked specificity because we couldn't query or filter by specific metadata fields that contained the text we wanted. In many cases, a context-specific R package is the best way to solve your data science needs. However, in this case, we found that using the NYT developer API was better suited to filtering and searching fields that contained the data most pertinent to our use case.

---

[7]https://cran.r-project.org/package=jstor
[8]https://cran.r-project.org/package=nasadata
[9]https://cran.r-project.org/package=rfacebookstat
[10]https://bradlindblad.github.io/schrute/
[11]https://en.wikipedia.org/wiki/Sentiment_analysis

7.6.1   Set Up Developer Account

Each user must have a developer account to access the NYT article database via API. Our first step is to access http://developer.nytimes.com and create an account. After creating your account, log in and select "Get Started" to begin the process of creating an API key, which is a code unique to you that authorizes access to the NYT articles database.

FIGURE 7.1 NYT Developer Network homepage.

7.6.2   Create an App

The first task after creating a developer task is to create an application and the unique database access key for that application to use. The steps to register an application (Network, 2022) are

1. Select **App** from the drop-down menu under your email address.
2. Click "**+ New App**" to create a new app.
3. Enter an App Name and Description for your app in the dialog box
4. Click **Create**.
5. Click the **APIs** tab.
6. Click the access toggle to **enable** access to the **Article Search API**.

In the last step, you'll notice several APIs to choose from, like Archive, Books, and Most Popular. In this case, you need a large text dataset, and the Article Search is the best fit. While the NYT API does not return full-text articles, we can select text-heavy metadata fields like abstracts in our API query that will meet our needs.

We'll need to look at the API menu within this site to get the specific syntax required to access the data we need. The Article Search API[12]

---

[12]https://developer.nytimes.com/docs/articlesearch-product/1/overview

page gives us the format for the API call specific to this dataset: /arti-
clesearch.json?q={query}&fq={filter}. There is a great deal of additional
information about this API that is useful, particularly how to access specific sections of the newspaper. This page also provide an example API call:
https://api.nytimes.com/svc/search/v2/articlesearch.json?q=election&api-
key=yourkey. When we compare the example call against the format for this
API is a specific web address plus our query terms and filters, as well as our
specific API key.

7.6.3 Find & Save Your API Key

App creation automatically triggers the creation of an API key. The key is
unique to your account and authorizes you to query the NYT database. Many
API calls you may utilize in your work will require an API key. To access your
API key,

1. Select **My Apps**.
2. Select your app.
3. View the API key on the **App Details** tab.
4. Confirm that the API key status is **Active**.

We need to add an object to store our API key so that our code can pass
that into the API call. Saving it as a separate object simplifies our code and
prevents us from inadvertently altering it when we update or change parts of
the API code itself.

To run this code chunk yourself, delete the #, copy your API key from the
App Details tab, and enter it between the double-quotes below.

```
# api_key <- "your_api_key_here"
```

7.6.4 Pull Data from the NYT API

Once the API key is stored, we are ready to construct the code to query
the NYT article database[13]. We will be using the *jsonlite* package to retrieve
the data in JSON format and then modify it into a form R can interpret.
JSON[14] is an acronym for JavaScript Object Notation and is a standard file

---

[13]the code in the section is a modification of code from https://rpubs.com/hmgeiger/373
949 and https://www.storybench.org/working-with-the-new-york-times-api-in-r/

[14]https://en.wikipedia.org/wiki/JSON

format used for transmitting data. It's programming-language agnostic, which means it can work with R, Python, and other languages. Text datasets are often encoded with JSON for export or transfer to other systems, making experience with the *jsonlite* package an instrumental skill.

We need three code sections to pull the data we need from the NYT. These code chunks will take us from our initial API query to a data frame of text field names and contents from the newspaper.

1. Identify the search query we want to pass to the database.

We want to search for the word "unemployment" in the US section of the NYT. This means our query will be "unemployment" and our filter will be the "U.S." section of the newspaper. We also have to append the API key to that search URL to authorize our API call. In the chunk below, `paste0()` is a base R function that concatenates two character strings. We want to join the search string with our API key for this search and save it as an object. Throughout this chapter, we will utilize this object, called `baseurl`, as a representation of our unique API call, which is the query string plus our specific API key.

```
baseurl <- paste0(
  "http://api.nytimes.com/svc/search/v2/articlesearch.json?
    q=unemployment&section_name=(U.S.)",
  api_key
)
```

2. Create an object to represent the number of search results pages we want to retrieve.

When using the API, each search results page contains ten records per page, and the API will only return one page of results at a time, no matter how many total pages the search returns. However, the ten records included on one page of search results will not provide enough data for text analysis so we will need many more than just one page. To balance the need for a large dataset without taking too long to process the data, we will retrieve output for 25 pages of search results. Before we run the query, we need to create an object that will store all of our API output.

```
pages <- vector("list", length = 25)
# 25 pages is 250 articles
```

3. Retrieve 25 pages of search results

We need to retrieve each page of search results, so that we can combine them into one dataset in a subsequent step.

```
page1 <- fromJSON(paste0(baseurl, "&page=1")) %>%
  data.frame()


page2 <- fromJSON(paste0(baseurl, "&page=2")) %>%
  data.frame()
```

Writing out the code for just two pages of search results is more than enough; who wants to write that code 23 more times and then concatenate them (c(page1, page2, ...)) together into one table?

Thankfully, as with the *purrr* package, R has options to handle the repetition for us. In this case, we want to run the same code 25 times, with only the page number changing. In other words, for every page between numbers 1 and 25, execute the API call and put the output into a single data frame. A particular code function called a **for-loop** makes this possible. For-loops have three parts (Wickham and Grolemund, 2016):

1. *output*, (here, the pages object)
2. *sequence*, the sequence to be looped over (here, 1-25)
3. *body*, (here, the baseurl plus the page number)

One characteristic of JSON data, which is used frequently in websites and web applications, is that the data table retrieved will have tables inside of tables. You could think of this as a row in a spreadsheet that points to another spreadsheet. We need to unpack these tables within tables using the FLATTEN parameter.

In the sequence below, i is a placeholder for each page number. We see this in two places, the for-loop as well in the output, where we want to create an object from each page (page i in pages 1-25). Also, there's a limit of 10 requests per minute for NYT API calls. Since we have 25 calls to make, we must pause for 6 seconds between page calls.

```
for (i in 1:25) {
  # can be read as "for every page number from 1 to 25"
  nytSearch <- fromJSON(paste0(baseurl, "&page=", i),
    flatten = TRUE
  ) %>% data.frame() # _body_
  pages[[i]] <- nytSearch # _output_
  Sys.sleep(6)
```

```
# there's a limit of 10 requests per minute for API calls
# we must pause for 6 seconds before initiating the next call
}
```

7.6.5   Create Textual Dataset from API Call

Our for-loop returned data that needs a little tidying before using it. We have
25 separate files with more metadata fields than we need.

1. First, we need to stitch each of the 25 pages of search results into
   one data frame.

```
articles_json <- rbind_pages(pages)
# combining 25 outputs into one dataframe
```

2. We need to pull out the four text-rich metadata fields for our analysis: abstract, snippet, lead paragraph, and headline. We need to
   view the metadata field names to ensure that we combine the correct
   column names.

```
colnames(articles_json)
```

```
##  [1] "status"
##  [2] "copyright"
##  [3] "response.docs.abstract"
##  [4] "response.docs.web_url"
##  [5] "response.docs.snippet"
##  [6] "response.docs.lead_paragraph"
##  [7] "response.docs.print_section"
##  [8] "response.docs.print_page"
##  [9] "response.docs.source"
## [10] "response.docs.multimedia"
## [11] "response.docs.keywords"
## [12] "response.docs.pub_date"
## [13] "response.docs.document_type"
## [14] "response.docs.news_desk"
## [15] "response.docs.section_name"
## [16] "response.docs.subsection_name"
```

```
## [17] "response.docs.type_of_material"
## [18] "response.docs._id"
## [19] "response.docs.word_count"
## [20] "response.docs.uri"
## [21] "response.docs.headline.main"
## [22] "response.docs.headline.kicker"
## [23] "response.docs.headline.content_kicker"
## [24] "response.docs.headline.print_headline"
## [25] "response.docs.headline.name"
## [26] "response.docs.headline.seo"
## [27] "response.docs.headline.sub"
## [28] "response.docs.byline.original"
## [29] "response.docs.byline.person"
## [30] "response.docs.byline.organization"
## [31] "response.meta.hits"
## [32] "response.meta.offset"
## [33] "response.meta.time"
## [34] "response.docs.slideshow_credits"
```

From the list of column names, we can see the specific field syntax we need.
Our code won't work if we don't have the exact column name as we start to
create a subset of the data for our use case.

3. We'll create a new object that is a data frame of only the four text
   columns we want to use.

```
text_fields <- articles_json %>%
  dplyr::select(c( # select() is a dplyr function
    response.docs.abstract, response.docs.snippet,
    response.docs.lead_paragraph, response.docs.headline.main
  ))
```

7.7 Tokenization

Just as tidy data principles dictate only one value per row/column pair, the
tidytext package requires a tidy textual dataset before analysis. We must convert our `text_fields` object from a row/column pair containing many sen-
tences into a table with only one word per row/column pair. Each word is

called a token, and breaking a text into individual tokens is called **tokeniza-tion**[15].

7.7.1   Use the tidytext Package to Load a Textual Dataset into the IDE and Tokenize It

We have four text fields, or columns, that we've pulled from the article database, and we need to tokenize each field. We will tokenize each column and then stitch them back together into a single, massive table for analysis in subsequent tasks.

   1. Tokenize abstracts

Using the `unnest_tokens()` function, we'll start to tokenize our dataset, beginning with the abstracts.

```
abstract_token <- text_fields %>%
  unnest_tokens(word, response.docs.abstract)

glimpse(abstract_token)
```

```
## Rows: 7,466
## Columns: 4
## $ response.docs.snippet         <chr> "The pandemic up~
## $ response.docs.lead_paragraph  <chr> "HONG KONG — Bef~
## $ response.docs.headline.main   <chr> "From Flight Att~
## $ word                          <chr> "the", "pandemic~
```

   2. Select only column 4 to create a clean table of abstract tokens

The `unnest_tokens()` function adds the tokens to the dataset as a new column. However, we only want the tokens column (the 4th column), so we must subset that column as we did in earlier chapters.

```
abstracts <- tibble(abstract_token[[4]])
```

Let's look at the first six results of the abstracts object.

---

[15]https://en.wikipedia.org/wiki/Lexical_analysis#Tokenization

```
head(abstracts)
```

```
## # A tibble: 6 x 1
##    `abstract_token[[4]]`
##    <chr>
## 1 the
## 2 pandemic
## 3 upended
## 4 careers
## 5 in
## 6 hong
```

3. Complete tokenization

We'll repeat the unnest_tokens() function for headlines, lead paragraphs, and snippets to finish tokenizing this dataset.

```
headline_token <- text_fields %>%
  unnest_tokens(word, response.docs.headline.main)

headlines <- tibble(headline_token[[4]])

lead_token <- text_fields %>%
  unnest_tokens(word, response.docs.lead_paragraph)

leads <- tibble(lead_token[[4]])

snippet_token <- text_fields %>%
  unnest_tokens(word, response.docs.snippet)

snippets <- tibble(snippet_token[[4]])
```

4. Combine tokens

After pulling the tokens out of their respective objects, we need to combine them into one very long column using c(). The single column in our table is unnamed, which we will rename so that we can use it in the next section, just like we did in Chapter 5.

```
words <- c(abstracts, headlines, leads, snippets) %>%
  unlist() %>%
  tibble() %>%
  rename(word = names(.))
```

7.8   Stop Words

When trying to intuit meaning from a textual dataset, we want to analyze the words that matter in the text and omit the insubstantial. Words like "of," "and," "to," and "by" don't convey meaning as "library," "classification," "books," and "reference" do. We call these words that have little value **stop words**[16], and we want to remove them from our dataset as part of the tidying process before we start analyzing the dataset. However, we cannot remove stop words until after tokenization. The stop words file has one word per row, so our dataset must also have one word per row. That way, the *dplyr* join function will match them.

As you remove stop words, however, be aware that what's considered a stop word can have implications for analysis in later steps. For example, "not" is used in English to negate another word, like "happy." Sentiment analysis is predicated on words having a positive or negative association, such as "happy" having a positive association. However, if "not" was used before "happy" and then removed as a stop word, that changes the negative textual meaning to a positive sentiment analysis.

7.8.1   Remove Stop Words from the Words Object

The *tidytext* package comes with a stop words dataset that we can load and run against our words dataset. We'll use a join from the *dplyr* chapter; in this case, we want to use anti_join(), which will remove any values that appear in both the words & stop_words datasets. This way, any word in the stop words list that appears in our dataset will be removed, along with stop words that have no match. An anti_join() must match on columns with the same name. The stop_words dataset has a column named word, and we renamed one of columns in the previous step to word also, so now our anti-join will work.

```
data("stop_words")
# loads the stop words set from tidytext

tidy_words <- words %>%
  anti_join(stop_words)
```

```
## Joining, by = "word"
```

```
# using a dplyr join, remove stop words found in the words dataset
```

---

[16]https://en.wikipedia.org/wiki/Stop_word

7.9   Sentiment Analysis

As previously mentioned, sentiment analysis is analyzing a text's feelings by rating the sentiment of each word in the text. To add objectivity to this analysis, data scientists utilize a lexicon to compare the text they're analyzing with a known list of word/sentiment pairs, like "happy"/positive. Bing, NRC, and the *textdata* package's lexicon (Mohammad and Turney, 2013) are several popular lexicons used in text mining.

Political scientists often use sentiment analysis to examine sentiment polarity, the positive/negative connotation of public discourse surrounding a political event (Matalon et al., 2021). Those studies often use social media datasets from Twitter or Facebook as a proxy for societal discourse at large. Similarly, law librarians may come across sentiment analyses of judicial opinions, the text of laws, or law review articles.

We are interested in the negative cultural connotation of the word "unemployment" and are looking for data to back that up. Now that we have a tidy, tokenized dataset, we can compare each term in our dataset against the Bing lexicon, which offers a binary positive/negative rating for each term in the lexicon. Using a *dplyr* join function, we can get the sentiment for each word in the dataset and calculate the positive and negative sentiment ratio.

7.9.1   Perform a Sentiment Analysis of the Textual Dataset

To determine the sentiment of our dataset of articles about unemployment, we will join the Bing lexicon (Hu and Liu, 2004) with our tokenized dataset and count the number of positive and negative emotions.

```
bing <- tidy_words %>%
  inner_join(get_sentiments("bing"))
```

```
## Joining, by = "word"
```

```
bing_count <- bing %>%
  count(sentiment)
```

```
print(bing_count)
```

```
## # A tibble: 2 x 2
##   sentiment     n
##   <chr>     <int>
## 1 negative    873
## 2 positive    410
```

This count shows that the 200 unemployment-related articles we analyzed are dominantly negative. By doing a little bit of calculation, we can see the exact ratio of negative to positive words.

```
ratio <- 877 / 378

print(ratio)
```

```
## [1] 2.32
```

Our data shows a rate of more than two negative terms for every one positive word. While we may intuitively know that unemployment has a negative connotation, we now have evidence to support that postulation. Being unemployed or looking for a job is hard enough without the specter of such a resoundingly negative cultural connotation. It would seem, then, that our outreach efforts are much needed.

7.10   TF-IDF

One of the critical underpinnings of text mining is the relationship between the frequency that a term appears in a collection of documents and how important or meaningful a term is to those documents. **Term frequency**, or tf, ranks every term in a text from most to least common. A term's frequency is a proportion created by dividing the number of terms in our dataset by counting how many times each word appears in the dataset. **Inverse document frequency**, or idf, relates to the premise that the most unique (least frequent) words are the most meaningful. This phenomenon is known as Zipf's Law[17], which says that the frequency of a word is inversely proportional to its rank.

When we're analyzing a textual dataset, using **tf-idf** can tell us the most meaningful words, which we can use to tell us the text's topic. With tf-idf, we give greater weight to terms that are essentially more unique in a particular text or corpus (group of many texts). Terms that are common in our texts but

---

[17]https://en.wikipedia.org/wiki/Zipf%27s_law

otherwise rare tell us what words are most important, and those words are most likely to indicate the subject matter. If we wanted to do a meta-analysis across many texts, we would then be able to compare and contrast datasets based on their topics. In this chapter, we will concern ourselves mostly with term frequency, but it is important to be familiar with tf-idf when doing textual analysis.

7.10.1 Determine Word Frequency

The first part of tf-idf is term frequency. To determine the frequency of the different words in our dataset, we need to count term frequency and sort them in descending order.

```
word_counts <- tidy_words %>%
  count(word, sort = TRUE)

head(word_counts)
```

```
## # A tibble: 6 x 2
##    word             n
##    <chr>        <int>
## 1 unemployment   749
## 2 percent        473
## 3 rate           356
## 4 jobs           143
## 5 labor          129
## 6 economy        105
```

This lists all the words in the dataset with a count of how many times each word appears, which gives us each word's frequency.

7.10.2 Identify the Text's Most Common Words

Once we have a count of each term's frequency in the overall dataset, we are able to perform the tf-idf calculation to determine which words are most important.

```
rank_frequency <- word_counts %>%
  # take our list of words, minus stop words
  mutate(
    rank = row_number(),
```

```
    # add a new column called rank (the row number for that row)
    term_frequency = n / 12731
)

print(rank_frequency)
```

```
## # A tibble: 2,768 x 4
##    word              n  rank term_frequency
##    <chr>         <int> <int>          <dbl>
##  1 unemployment    749     1         0.0588
##  2 percent         473     2         0.0372
##  3 rate            356     3         0.0280
##  4 jobs            143     4         0.0112
##  5 labor           129     5         0.0101
##  6 economy         105     6        0.00825
##  7 jobless          96     7        0.00754
##  8 time             95     8        0.00746
##  9 department       83     9        0.00652
## 10 job              82    10        0.00644
## # ... with 2,758 more rows
```

The new column term_frequency is a calculated field because it is derived from
the results of a mathematical operation. We sorted each word by the number
(n) of times it appears in our text. There were a total of 12731 observations
(rows) in tidy_words. The result is a ratio of how many times each word
appeared in the text compared with the number of words in total.

7.10.3 Plot Word Frequency

Making a plot to show term frequencies is a way to ensure that the textual
dataset has a normal distribution. Due to Zipf's Law, a "normal" distribution
here would be a constant negative slope; the straight line should be higher on
the left and lower on the right. However, when we look at the term_frequency
column in the preceding step, we notice that the numbers start small and
then get progresively smaller. If we made a plot of this data as-is, it would
not follow the bell-curve of what statisticians call a "normal distribution." In
cases like these, a data scientist might want to transform the entire dataset by
taking the $\log(x)$ of each value. We'll apply a logarithmic scale to both our x
and y axis with the scale_*_log10() functions in our ggplot() function call:

```
rank_frequency %>%
  ggplot(aes(rank, term_frequency, color = word)) +
```

```
geom_line(
    group = 1,
    size = 1.1,
    alpha = 0.8,
    show.legend = FALSE
) +
scale_x_log10() +
scale_y_log10()
```

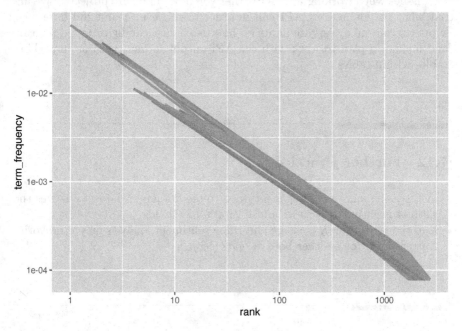

```
# the logarithmic scale on both axis shows the constant negative slope
```

Plotted above, we can see that our textual dataset conforms to Zipf's Law, which validates the normalcy of our dataset and the other conclusions we have drawn about it. By transforming our original skewed, long-tail dataset we were able to create a linear relationship and display the constant negative slope. This highlights the inverse relationship between word frequency and meaning.

7.11   Summary

This chapter covered the basics of what text mining, sentiment analysis, and tf-idf are. We started with APIs and JSON to extract data from a popular article database and then walked through the steps involved in tidying and performing a text analysis on that data. We covered several structural planning issues when working with APIs that you'll use in future projects: figuring out what to put in your API call in order to retrieve meaningful data that supports your analysis, rate limits for how much data can be returned per call, and making a plan to work with the API output format, such as unpacking tables within tables.

7.12   Further Practice

- increase the number of pages returned to see if a larger dataset changes the ratio of negative to positive and if Zipft's Law holds
- load the *janeaustenR* package and run a sentiment analysis on your favorite Austen novel (or another book in *gutenbergR*)

7.13   Additional Resources

- *Text Mining in R*: https://www.tidytextmining.com/
- Digital Humanities resources:
  - earlyprint.org: "The EarlyPrint Lab offers a range of tools for the computational exploration and analysis of English print culture before 1700"
  - americaspublicbible.org

8

Creating Dynamic Documents with rmarkdown

8.1 Learning Objectives

1. Reproduce markdown syntax.
2. Use appropriate code to load data and relevant libraries in RStudio.
3. Generate code to create and format code chunks.
4. Modify display options to create a customized R Markdown document.
5. "Knit" R Markdown documents into various formats.

8.2 Scenario

Now that we have a foundation, let's use everything we've learned so far to create an R Markdown document with our analyses of the US Census and NYT data, as well as our data visualizations (the plots and maps we created earlier) for the relevant stakeholders.

8.3 Introduction to R Markdown

The *rmarkdown* package allows you to create various documents such as PDFs, Microsoft Word documents, and HTML files. You can also create websites, presentations, and dashboards. You will build on your R Markdown knowledge to create a dashboard in Chapter 9. R Markdown uses Markdown syntax[1] to add

---

[1] https://www.markdownguide.org/getting-started/

DOI: 10.1201/9781003218012-8

formatting to plain text documents. Examples of such formatting are: adding headers, italicizing and bolding text, and adding ordered and unordered lists. We are not doing a deep dive with R Markdown, but by the time you finish this chapter, you will be up and running with R Markdown in creating a basic document. For more information about R Markdown, please refer to *R Markdown the Definitive Guide*, by Yihui Xie, J. J. Allaire, and Garrett Grolemund[2].

In addition, R Markdown allows you to include executable code chunks within the R Markdown document. So what's the point in putting in code chunks within an R Markdown document when you can give someone a file with your code? Why not use Microsoft PowerPoint and Word instead? Having the code chunks embedded in the document can serve various purposes. Suppose you want to create a document or presentation with data visualizations such as bar plots and an explanation of the visualization. In that case, you can do that all within one ecosystem instead of creating the visualization in another program, saving the image, and then adding it to your document or presentation. Also, suppose you have to produce regular reports that use continuously updated datasets. In that case, you can easily update your visualization by merely reloading your updated dataset and publishing it. This is transformative for workflows that entail running the same data analysis on a monthly or annual basis.

8.4   Packages & Datasets Needed

For this chapter, you will need to install and load the *tidyverse, tidycensus, readr, sf,* and *tmap* packages. To create the R Markdown document, we also need to load the necessary data if it isn't already loaded. We will be loading the CSV files showing the occupations with the lowest unemployment by sex and our shapefiles of the St. Louis wards and Census tracts, which we created in earlier chapters.

```
library(tidyverse)
library(tidycensus)
library(readr)
library(sf)
library(tmap)

# csv files
male_low_unemployment <- read_csv("data/male-low-unemployment.csv")
female_low_unemployment <- read_csv("data/female-low-unemployment.csv")

# shapefiles
```

---

[2]https://bookdown.org/yihui/rmarkdown/

```
stl_wards <- st_read("nbrhds_wards/WARDS_2010.shp")
unemployment_tract <- st_read(
  "unemployment_tract/unemployment_tract.shp"
)
```

8.5   Creating an R Markdown Document

You can create an R Markdown document in RStudio by going to File > New File > R Markdown. You can create a document, presentation, Shiny app, or you can use a specific template. Keep the defaults selected and click OK.

8.6   R Markdown Document Structure

Now that we loaded the necessary libraries and data, it's time to talk about the R Markdown document structure. R Markdown documents consist of three components:

1. YAML header

2. R Markdown syntax

3. Code chunks

8.6.1   YAML header

YAML, or "YAML Ain't Markdown Language," is used in the header to set the parameters. It contains the metadata on the document. Here is the default YAML for an R Markdown document:

```
---
title: "Enter title here"
author: "Author's Name"
date: "Date"
output: html_document
---
```

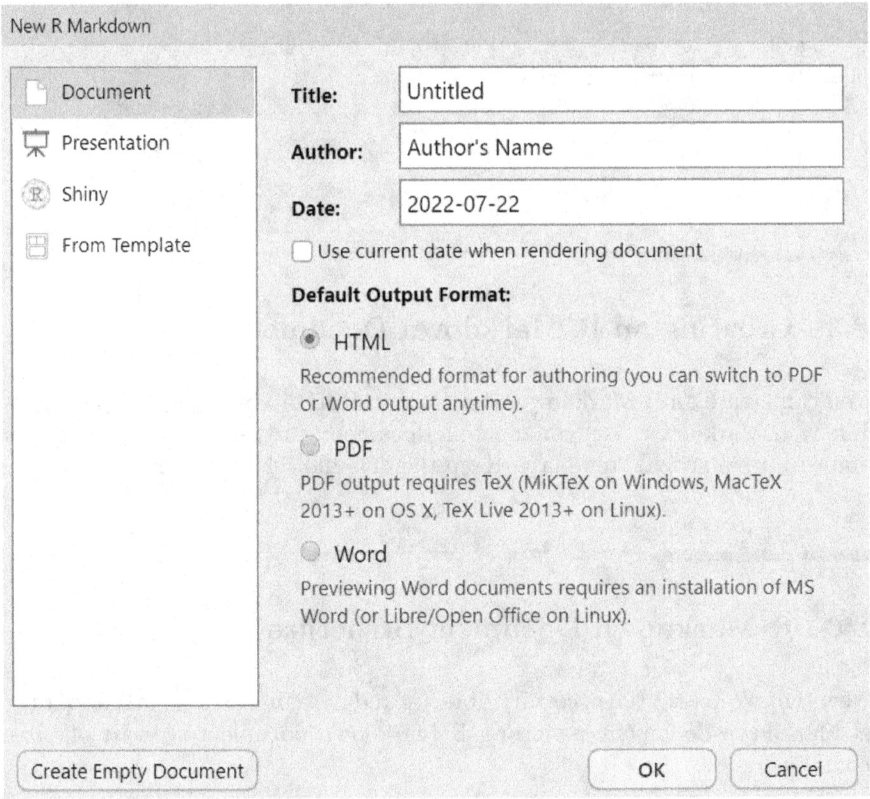

FIGURE 8.1 R Markdown document creation prompt.

This default YAML indicates the title, author, date, and output type. There are a variety of output options that you can designate in the YAML header. For example, in the above code, we specified the output as an HTML document. There are various output options, such as a PDF, Word, or PowerPoint

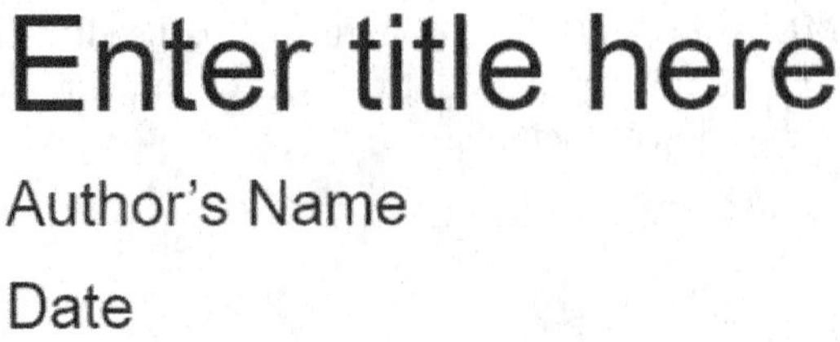

FIGURE 8.2 Metadata information.

presentation[3]. You can also customize your R Markdown document using templates which we will discuss later in this chapter.

8.6.2 R Markdown Syntax

The R Markdown syntax allows you to add specific styling to your document, such as adding headers, italicizing and bolding texts, adding ordered and unordered lists, and adding links.

Creating headers

- # is header 1
- ## is header 2
- ### is header 3

Other styles

- * Before and after the words: Italicized
- **Before and after the words: Bold

Creating lists

- - after each item on the list, create an unordered list.
- 1. and continuing the numbers sequentially to create an ordered list.

Links

- [Website Link](www.website.com): the bracket contains the link text while we put the web address within the parentheses.

8.6.3 Embedding Code

You can embed code in your R Markdown document by adding code chunks. You can do this by clicking on the green "C" icon at the top of the Editing pane, and you can choose to add a code chunk for a variety of coding languages. You can also manually add the code chunk by typing the code below.

---

[3]https://bookdown.org/yihui/rmarkdown/output-formats.html

Creating headers

- # is header 1
- ## is header 2
- ### is header 3

Other styling - * Before and after the words: Italicized - ** Before and after the words: Bold

Creating lists - after each item on the list to create an unordered list. 1. and continuing the numbers sequentially to create an ordered list.

Links [Website Link](www.website.com) The bracket contains the link text while the web address is put within the parentheses.

FIGURE 8.3 Showing the output for headers, another styling like italicization and bolding along with unordered and ordered lists.

Code chunk options

You can set various options related to the code output and the display of the code within {r} section of the code chunk.

- include = FALSE prevents the code output and code from being shown in the document you create. Other chunks can still use the output of the code.
- echo = FALSE shows the output but not the code.

Let's show how this works with loading the information about the aldermen.

Code example with echo = FALSE

```
## Rows: 29 Columns: 4
## -- Column specification ------------------------------------------
-----
## Delimiter: ","
## chr (4): Ward, Alderman, website, Email
##
## i Use `spec()` to retrieve the full column specification for this data.
## i Specify the column types or set `show_col_types = FALSE` to quiet this message.
```

Code example with echo = TRUE

```
#code example with using echo as TRUE
aldermen_info = read_csv("data/aldermen-contact.csv")
```

```
## Rows: 29 Columns: 4
## -- Column specification ------------------------------------------
-----
## Delimiter: ","
## chr (4): Ward, Alderman, website, Email
##
## i Use `spec()` to retrieve the full column specification for this data.
## i Specify the column types or set `show_col_types = FALSE` to quiet this message.
```

With `echo = FALSE`, you will not see the code in the output of reading in the csv with the alderman contact information, while when `echo = TRUE`, you can see the code and the output.

Code example with `include = TRUE`

```
# code example with using include = TRUE
aldermen_info = read_csv("data/aldermen-contact.csv")
```

```
## Rows: 29 Columns: 4
## -- Column specification ------------------------------------------
-----
## Delimiter: ","
## chr (4): Ward, Alderman, website, Email
##
## i Use `spec()` to retrieve the full column specification for this data.
## i Specify the column types or set `show_col_types = FALSE` to quiet this message.
```

Code example with `include = FALSE`

When loading the alderman info CSV with `include = TRUE`, you will see both the output and the code, but with `include = FALSE`, you will not see either the output or the code.

8.7 Adding Data Visualizations to Your R Markdown Document

You can also embed data visualizations such as plots and maps into R Markdown. Let's embed the plot we made in the *ggplot* chapter in our R Markdown document. We added a code chunk and made sure `echo = TRUE` to see the code output.

Plot of the top ten female occupations in Census tracts with lowest unemployment

```
low_unemployment_female_plot <- female_low_unemployment %>%
  ggplot(aes(x = total, y = reorder(female_jobs, total))) +
  geom_col() +
  labs(
```

```
    title = "Top 10 female jobs in areas with low unemployment",
    x = "Total",
    y = "Occupation"
)
```

```
low_unemployment_female_plot
```

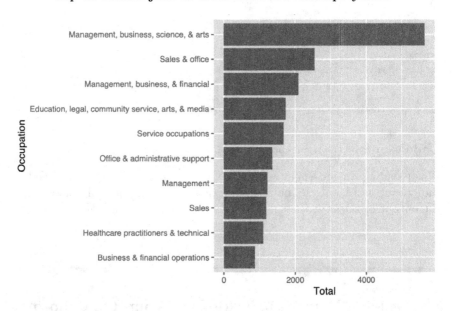

Top 10 female jobs in areas with low unemployment

You can also embed maps into your R Markdown document. As with the code chunk before, we make sure the setting is `echo = TRUE` so we can see the output of the code, which will be a map of the wards in St. Louis.

```
tm_shape(stl_wards, unit = "mi") +
  tm_polygons() +
  tm_layout(
    title = "Wards in St. Louis City",
    title.size = .7
  ) +
  tm_scale_bar(position = c("right", "bottom"), width = .3) +
  tm_compass(position = c("left", "top"))
```

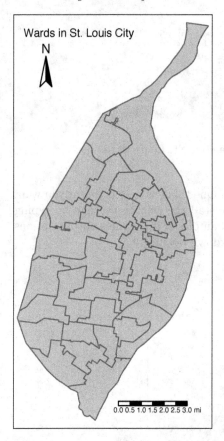

8.8 Creating Your Output

You know how to add stylistic elements and code to your R Markdown document, but how do you transform the R Markdown syntax and chunks in your R Markdown file to an output document like a PDF or Word file? Rendering is the process of converting the R Markdown syntax and code chunks into a document. The *knitr* package is required to do this rendering. There are two ways in which you can do the rendering. When you create an R Markdown document in RStudio, you will see a Knit button in the document toolbar. If you press the arrow next to the Knit icon, you will see options to Knit to

You can also embed maps into your R Markdown document. As with the code chunk before, we make sure the setting is echo = TRUE so we can see the output of the code, which will be a map of the wards in St. Louis.

```
tm_shape(stl_wards, unit = "mi")+
  tm_polygons()+
  tm_layout(title = "Wards in St. Louis City", title.size = .7) +
  tm_scale_bar(position = c("right", "bottom"), width = .3)+
  tm_compass(position = c("left", "top"))
```

FIGURE 8.4 PDF output showing the code St. Louis wards map.

PDF, Knit to HTML, and Knit to Word. One thing to note is that if you want to create a PDF document, you must install the *tinytex* package by typing tinytex::install_tinytex(). Here are examples of the various output options displaying the same content in the markdown document.

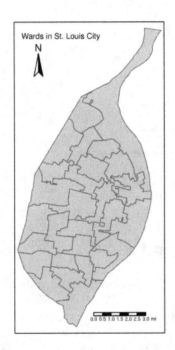

FIGURE 8.5 PDF output showing the map of St. Louis wards.

You can also embed maps into your R Markdown document. As with the code chunk before, we make sure the setting is `echo = TRUE` so we can see the output of the code, which will be a map of the wards in St. Louis

```
tm_shape(stl_wards, unit = "mi")+
  tm_polygons()+
  tm_layout(title = "Wards in St. Louis City", title.size = .7) +
  tm_scale_bar(position = c("right", "bottom"), width = .3)+
  tm_compass(position = c("left", "top"))
```

FIGURE 8.6 HTML document showing a map of St. Louis wards and the code that created the map.

8.9 Customizing Your R Markdown Document with Templates

You can customize your R Markdown document through a template you create on your own or through a pre-built template and call it through the "output" parameter in the YAML headers. We will focus on using a pre-built template. You can look at examples of R Markdown templates on the R Markdown gallery[4]. There are existing templates you can choose from when you create a new R Markdown document, and you can also add a template by changing the output parameter of the YAML header.

---

[4]https://rmarkdown.rstudio.com/gallery.html

You can also embed maps into your R Markdown document. As with the code chunk before, we make sure the setting is echo = TRUE so we can see the output of the code, which will be a map of the wards in St. Louis.

```
tm_shape(stl_wards, unit = "mi")+
  tm_polygons()+
  tm_layout(title = "Wards in St. Louis City", title.size = .7) +
  tm_scale_bar(position = c("right", "bottom"), width = .3)+
  tm_compass(position = c("left", "top"))
```

FIGURE 8.7 Word document showing a map of St. Louis wards and the code that created the map.

Let's add one of the prettydoc[5] templates called "architect" to your R Markdown document.

```
---
title: "Report title"
author: "Your Name"
date: July 31, 2016
output:
 prettydoc::html_pretty:
    theme: cayman
    highlight: github
---
```

---

[5] https://prettydoc.statr.me/themes.html

Report title

Your Name

July 31, 2016

```
#Loading packages
library(tidyverse)
```

```
## -- Attaching packages ---------------------------------------- tidyverse 1.3.1 --
```

```
## v ggplot2 3.3.5      v purrr   0.3.4
## v tibble  3.1.6      v dplyr   1.0.7
## v tidyr   1.2.0      v stringr 1.4.0
## v readr   2.1.2      v forcats 0.5.1
```

```
## -- Conflicts ------------------------------------------------ tidyverse_conflicts() --
## x dplyr::filter() masks stats::filter()
## x dplyr::lag()    masks stats::lag()
```

```
library(tidycensus)
library(readr)
library(sf)
```

You can also embed maps into your R Markdown document. As with the code chunk before, we make sure the setting is `echo = TRUE` so we can see the output of the code, which will be a map of the wards in St. Louis.

```
tm_shape(stl_wards, unit = "mi")+
    tm_polygons()+
    tm_layout(title = "Wards in St. Louis City", title.size = .7) +
    tm_scale_bar(position = c("right", "bottom"), width = .3)+
    tm_compass(position = c("left", "top"))
```

8.10 Summary

R Markdown is a way to create various types of documents in RStudio. Using the Markdown language, it can be used to format plain text documents. An R Markdown document comprises a YAML header containing the metadata, R Markdown syntax allowing you to add stylistic elements in the document, and code chunks. When adding code chunks, you can choose whether to display the code chunks along with the output. After creating your R Markdown document, you can create a Word, PDF, or HTML document through knitting. Knitting converts R Markdown into the output format designated in the YAML header.

8.11 Resources

- The R Markdown website: https://rmarkdown.rstudio.com
- *R Markdown, The Definitive Guide*: https://bookdown.org/yihui/rmarkdo wn/

8.12 Further Practice

- Create an R Markdown document and add a prettydoc template. Try changing the default YAML output to the other two prettydoc templates, which are `cayman` and `tacticle`.
- Posit PBC staff have created a new publishing format that works with many programming languages in addition to R, called Quarto. Using the documentation at https://quarto.org/, install the *quarto* package[6] and recreate your R Markdown document using Quarto.

---

[6]https://cran.r-project.org/web/packages/quarto/index.html

9

Creating a Flexdashboard

9.1 Learning Objectives

1. Define a flexdashboard and the basic functionality of a flexdashboard.
2. Generate an R Markdown document with flexdashboard as the output format.
3. Recognize various layout structures of the flexdashboard in terms of column and row orientation.
4. Create a flexdashboard with a provided layout.

9.2 Terms You'll Learn

- Flexdashboard

9.3 Scenario

Now that you know how to do various tasks with R such as data scrubbing and creating different data visualizations, you need to create a dashboard so the stakeholders can be able to easily access the information you've presented. You will use the *flexdashboard* package to do so. Your final product should look like this example from rpubs.com[1]: https://rpubs.com/sarahemlin/stl-dashboard.

Note: You should create a blank R-Markdown document when creating this dashboard because the elements of the dashboard will not run within the chapter.

---

[1]RPubs is a free hosting platform for R Markdown documents provided by Posit, PBC

DOI: 10.1201/9781003218012-9

FIGURE 9.1 St. Louis dashboard on RPubs.

9.4   Overview of flexdashboard

Flexdashboard allows you to create a dashboard using R Markdown. It is a flexible interface that allows you to create a dashboard of various data visualizations into one interface. The main components of a flexdashboard consist of pages, columns, and rows. Within these components, you can insert various elements such as tabular output, graphical output, or interactive data visualizations that run under the JavaScript programming language[2].

9.5   Packages and Datasets Needed

We are going to load several familiar packages such as *tidyverse*, *tmap*, and *sf*, with the addition of two new packages which are *DT* and *flexdashboard*.

```
library(tidyverse)
library(flexdashboard)
library(tmap)
library(sf)
library(DT)
```

---

[2]https://pkgs.rstudio.com/flexdashboard/index.html

We are also going to load the data that we used in the previous chapters. This includes the CSV files of the alderman contact information and the male and female unemployment information, along with the shapefiles of the St. Louis wards and Census tracts. We also need to do a simple data transformation to calculate the unemployment rate using the *dplyr* `rename()` and `mutate()` functions.

9.6 Initiating the flexdashboard

There are two ways in which you can initiate a flexdashboard. One way is to create it from the File tab in RStudio by going to File > New File > R Markdown > From Template.

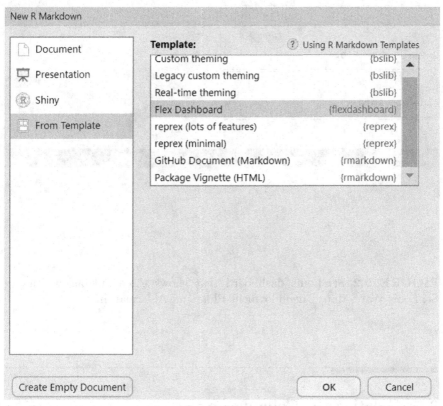

Another way we can create a flexdashboard is to create an R Markdown document and then specify the flexdashboard as output in the YAML header.

```
---
title: "St. Louis Dashboard"
output: flexdashboard::flex_dashboard
---
```

9.7   Creating the Pages

First, we will create three pages in our dashboard: the St. Louis ward maps, unemployment plots, and alderman information.

```
St. Louis Ward Map
===================================


Unemployment Plots
===================================

Alderman Info
===================================
```

FIGURE 9.2 St. Louis dashboard that shows three buttons which are St. Louis Ward Map, Unemployment Plots and Alderman Info.

9.8   Creating the Columns

To do an easy comparison, we will place the maps in the **St. Louis Ward Map** page and the plots in the **Unemployment Plots** page side by side.

We will do this by adding columns to both these pages. We will not add any columns to the **Aldermen Info** page.

```
St. Louis Ward Map
===================================


Column 1
-------------------------------------
St. Louis Wards



Column 2
-------------------------------------
Unemployment Rate in St. Louis by Ward



Unemployment Plots
===================================


Column 1
-------------------------------------

### Lowest unemployment (female)


Column 2
-------------------------------------
### Lowest unemployment(male)



Alderman Info
===================================
```

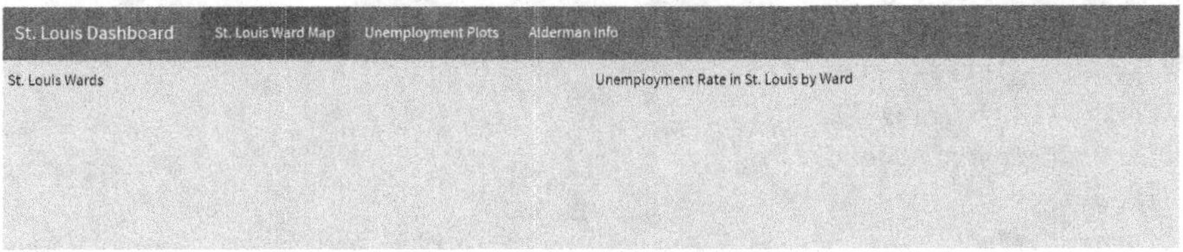

FIGURE 9.3 St. Louis Wards Map page with two columns.

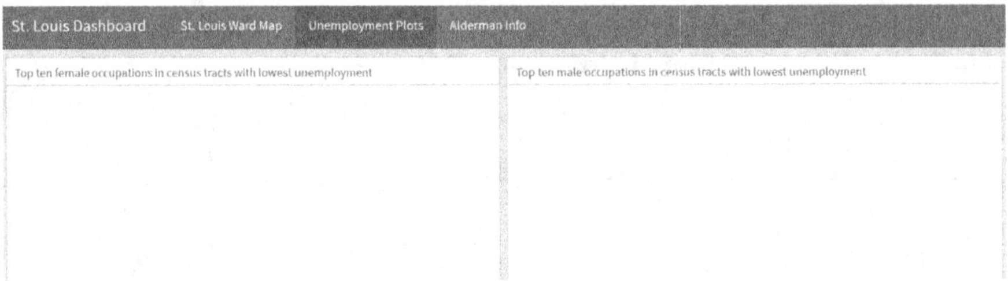

FIGURE 9.4 Unemployment plots page with two columns.

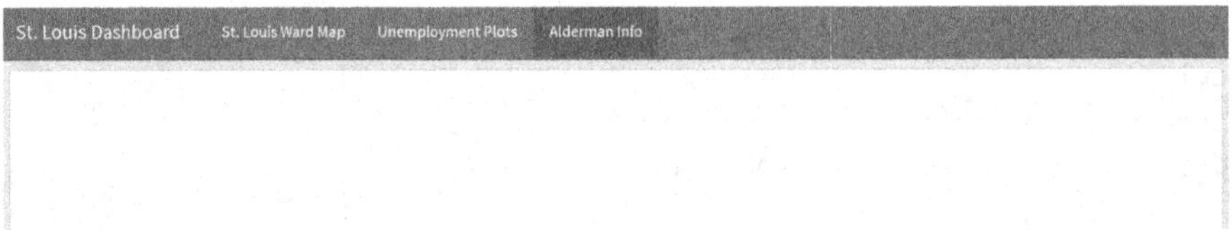

FIGURE 9.5 Aldermen Info page.

9.9   Adding the Code Chunks

We will then put the title of the column sections and code chunks below these column designations. For the Alderman info page, we are using the `datatable` function to display the table of alderman contact information. We will remove the titles of the unemployment plots added with *ggplot2* since we already have a column title for each plot.

```
---
title: "St. Louis Dashboard"
output: flexdashboard::flex_dashboard
---


```{r setup, include=FALSE}
knitr::opts_chunk$set(echo = TRUE)
# loading libraries
library(tidyverse)
library(flexdashboard)
library(tmap)
library(sf)
library(DT)

```
```

```{r, include = FALSE}
aldermen_info = read_csv("data/aldermen-contact.csv")
male_low_unemployment <- read_csv("data/male-low-unemployment.csv")
female_low_unemployment <- read_csv("data/female-low-unemployment.csv")
stl_wards <- st_read("nbrhds_wards/WARDS_2010.shp")
stl_tracts <- st_read("unemployment_tract/unemployment_tract.shp")%>%
  rename("unemployment_rate" = "unmply_") %>%
  mutate(unemployment_rate = unemployment_rate * 100)
```

St. Louis Ward Map
===================================

Column 1

St. Louis Wards

```{r ward-map, echo=FALSE}
tmap_mode("view")
tm_shape(stl_wards, unit = "mi")+
  tm_polygons()

```

Column 2

Unemployment Rate in St. Louis by Ward

```{r unemployment-map, echo = FALSE}
tmap_mode("view")
tm_shape(stl_tracts) +
  tm_fill("unemployment_rate", title= "Umemployment rate",
          popup.vars = c("% unemployed" = "unemployment_rate"),
          id = "NAME")

```

Unemployment Plots
===================================

Column 1

```
-----------------------------------

**Lowest unemployment (female)**
```{r plot-female, echo= TRUE}
low_unemployment_female_plot <- female_low_unemployment %>%
  ggplot(aes(x = total, y = reorder(female_jobs, total))) +
  geom_col() +
  labs(x = "Total", y = "Occupation")

low_unemployment_female_plot
```


Column 2

Lowest unemployment(male)
```{r plot-male, echo= TRUE}
low_unemployment_male_plot <- male_low_unemployment %>%
  ggplot(aes(x = total, y = reorder(male_jobs, total))) +
  geom_col() +
  labs(x = "Total", y = "Occupation")

low_unemployment_male_plot
```


Alderman Info
===================================
```{r alderman, echo = TRUE}
datatable(aldermen_info)
```
```

9.10 Summary

You can use R Markdown to create dashboards with the *flexdashboard* package. With the *flexdashboard* package, we can create dashboards that have
different pages and we can add columns to those pages. In this chapter, we
created a St. Louis Dashboard with three pages for the St. Louis ward maps,
unemployment plots, and aldermen information.

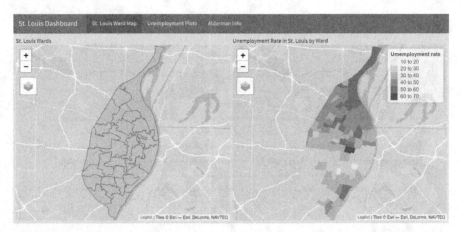

FIGURE 9.6 St. Louis Ward Maps page.

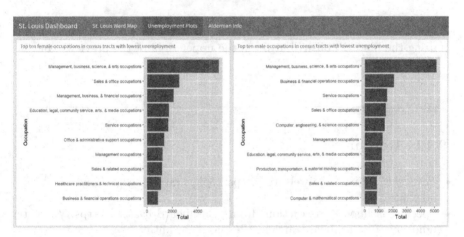

FIGURE 9.7 Unemployment plots page.

9.11   Further Practice

- You can apply various themes[3] to your flexdashboard. Try to apply a theme to the St. Louis Dashboard.
- You will need to add a page that has an R markdown document summarizing your findings. Create another tab called "Report" and put your summary in this section.

---

[3]https://rstudio.github.io/flexdashboard/articles/theme.html

FIGURE 9.8 Alderman info page.

- Create a free account on https://rpubs.com and publish your flexdashboard using the blue "Publish" button in the upper right of the preview pane.

9.12   Resources

- *flexdashboard* package website: https://pkgs.rstudio.com/flexdashboard/index.html
- Specific dashboard instructions from "Dashboard basics": https://pkgs.rstudio.com/flexdashboard/articles/flexdashboard.html

10

Creating an Interactive Dashboard with Shiny

10.1 Learning Objectives

1. List the basic components of a Shiny app.
2. Explain the functions of the user interface and the server.
3. Describe reactivity in Shiny apps.
4. Generate code in the Shiny user interface to access the provided data.
5. Generate code to in the Shiny server function to access the provided data.

10.2 Terms You'll Learn

- Reactivity

10.3 Scenario

Now that you created the flexdashboard, you want to have some interactivity in your dashboard such as the users being able to provide input which would change the plots or the maps. You heard that you would be able to do that using the *shiny* package.

DOI: 10.1201/9781003218012-10

10.4   Packages and datasets needed

We will be using the same packages and data from the chapter on flexdash-board with the addition of *shiny*.

```
# loading libraries
library(tidyverse)
library(flexdashboard)
library(tmap)
library(sf)
library(DT)
library(shiny)
```

10.5   What You Need to Know About Shiny

The purpose of this chapter is to get you up and running with using *shiny* by introducing you to the most essential concepts of a Shiny web application and the basic components. By the end of this chapter, you will be able to develop a simple Shiny app that is interactive between tabular data and data visualizations. This is in no way a thorough introduction to Shiny. For a more extensive explanation, please read *Mastering Shiny*[1] by Hadley Wickham.

In short, *shiny* is a package that allows you to create interactive web applications For example, you can create a simple web application of an interactive plot that can change by what the users select with a dropdown menu. One example of this is the Telephones by Region[2] app on the online Shiny Gallery.

Shiny apps are composed of three parts, which are the user interface (UI), the server, and the shinyApp function. The UI is what the user will see on the front end and it is where you build the look of the app. In the UI, you can add various elements that will allow the user to interact with the app such as buttons, sliders, and drop down menus along with the display of your data visualization or data table. On the back end, the server controls what the app will do. For example, it handles what happens when a user interacts with the app such as pressing a button. The interaction between the UI and server is based on **reactivity** in which the outputs update based on the input

---

[1] https://mastering-shiny.org/basic-app.html
[2] https://shiny.rstudio.com/gallery/telephones-by-region.html

(Wickham, 2021). This concept of reactivity will become more clear as we build our app. To put these both together, so you can deploy your app you need to call the `shinyApp` function. Let's first create shiny apps of each component of the dashboard, and then we will integrate these apps in the flexdashboard that you created.

10.6 Creating a Shiny Web Map App

10.6.1 Initalizing from RStudio

In RStudio, you can create a Shiny app by going to File > Shiny Web App. You will be prompted to give an application name and application type. Name the app "stl_web_map" and select the Single File (app.R) type since we are creating a small app. The Multiple File(ui.R/server.R) type is best when you are creating larger shiny apps[3]. To get started, we need to create the user interface and the server functions.

```
# creating the ui
ui <- fluidPage()

# creating the server
server <- function(input, output) {
}

# running the app
shinyApp(ui = ui, server = server)
```

Initially, the output will be blank since we haven't added any elements. One we've started, we need to add a dropdown menu which will allow us to select what layers we want to see on the map. First, create a variable called "layers" which will contain the dropdown menu choices which are "Select a variable," "Unemployment rate by ward," and "St. Louis Wards." After that, call the `selectInput` function which contains the name in which we will refer the dropdown menu when we need to call it in the server, the title of the dropdown menu bar, along with our choices. By default, the first option, "Select a variable" is selected.

---

[3]https://shiny.rstudio.com/articles/two-file.html

```r
layers <- c(
  "Select a variable",
  "Unemployment rate by ward",
  "St. Louis Wards"
)

# creating the ui
ui <- fluidPage(
  selectInput(
    "var",
    "St. Louis Maps",
    choices = layers,
    selected = layers[1]
  )
)

# creating the server
server <- function(input, output) {
}

# running the app
shinyApp(ui = ui, server = server)
```

St. Louis Maps

Select a variable   ▼

FIGURE 10.1 Select input bar.

Next, we will create the back-end functionality of the app in the server. First, let's add code which will display, or render, the map. We do that by using the `tmapOutput` function and giving that output a name which we can refer to when creating the reactive variables. After the `tmapOutput` function, we define the reactive variable `output$map` which allows the map to be instantly updated based on the user's input. Since we are rendering a map based on the *tmap* package, we will be using the `renderTmap` function. Some of the following code should be familiar: we are creating a *leaflet* map which will build the map based on the St. Louis ward polygons. The one addition is setting the z-index in `tm_polygons`. A z-index specifies the order of overlapping HTML elements. The z-index is set to the layer number plus 400[4].

---

[4]https://cran.r-project.org/web/packages/tmap/tmap.pdf

```r
layers <- c(
  "Select a variable",
  "Unemployment rate by ward",
  "St. Louis Wards"
)

# creating the ui
ui <- fluidPage(
  selectInput(
    "var",
    "St. Louis Maps",
    choices = layers,
    selected = layers[1]
  )
)

# creating the server
server <- function(input, output) {
  # this is the map that will load
  tmapOutput("map")

  output$map <- renderTmap({
    tmap_mode("view")
    tm_shape(data, unit = "mi") +
      tm_polygons(zindex = 401)
  })
}

# running the app
shinyApp(ui = ui, server = server)
```

The next thing we are going to do is add reactivity to the app by adding the observeEvent function. We want the app to update based on the selection in the drop-down menu. The observeEvent function will run a segment of code based on the reactive variable that is selected which will create the output (Wickham, 2021). We also add if statements to complement the observeEvent function, in order to run certain code chunks based on a specific condition (selected in the dropdown menu). A good way to remember reactivity is that reactive variables are dynamic inputs, while observers are dynamic outputs.

```r
layers <- c(
  "Select a variable",
```

St. Louis Maps

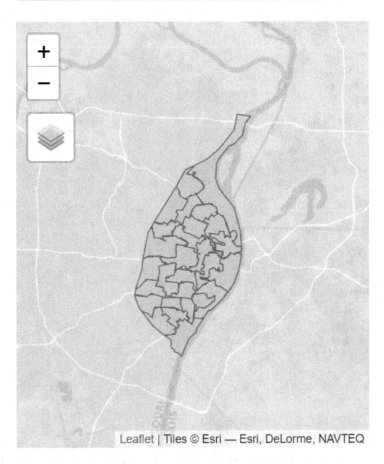

FIGURE 10.2 Shiny app showing the St. Louis ward map.

```
  "Unemployment rate by ward",
  "St. Louis Wards"
)

# creating the ui
ui <- fluidPage(
  selectInput(
    "var",
```

```
      "St. Louis Maps",
      choices = layers,
      selected = layers[1]
  )
)


# creating the server
server <- function(input, output) {
  # this is the map that will load
  tmapOutput("map")

  output$map <- renderTmap({
    tmap_mode("view")
    tm_shape(data, unit = "mi") +
      tm_polygons(zindex = 401)
  })


  observeEvent(input$var, {
    if (input$var == layers[2]) {
      data <- stl_tracts
      tmapProxy("map", session, {
        tm_remove_layer(401) +
          tm_shape(data) +
          tm_fill("unemployment_rate",
            title = "Umemployment rate",
            popup.vars = c("% unemployed" = "unemployment_rate"),
            id = "NAME"
          ) +
          tm_polygons(zindex = 401)
      })
    }


    if (input$var == layers[3]) {
      data <- stl_wards
      tmapProxy("map", session, {
        tm_remove_layer(401) +
          tm_shape(data) +
          tm_polygons(zindex = 401)
      })
    }
  })
}
```

```
# running the app
shinyApp(ui = ui, server = server)
```

Let's break down the observeEvent function. As mentioned before, the ob-serveEvent function will change the output based on the input. Each visu-alization has a different way of doing this. In the case of *tmap*, the tmap-Proxy function will update the map once various conditions are met. If the user selects the "Unemployment rate by ward" selection (layers[2]), then the tmapProxy function will delete the previously displayed layer and add the "Un-employment rate by ward" layer. If the user selects the "St. Louis Wards" (layers[3]) option, then the tmapProxy will update the map to show that layer. The way the layer is removed is through calling the z-index of the layer that is being displayed, which is 401 in this case.

We are done creating this app! Run it to see what it looks like.

10.7   Creating a `ggplot` Shiny App

Now that we have created the Shiny mapping app, let's go ahead and create an interactive plotting app. Make sure to create a new file to create the new Shiny app by going to File > New File > Shiny Web App.

We are not going to go step by step with this app, because the steps are the same as above except for two substitutions: we will use a different output function to display the visualization (plotOutput()) and a different render function to display the plot (renderPlot()).

```
stl_plot_choices <- c(
  "Choose a plot",
  "Occupations with the lowest unemployment (female)",
  "Occupations with the lowest unemployment (male)"
)

# creating the ui
ui <- fluidPage(
  selectInput(
    "plots",
    "Plots",
    choices = stl_plot_choices,
```

St. Louis Maps

St. Louis Wards   ▼

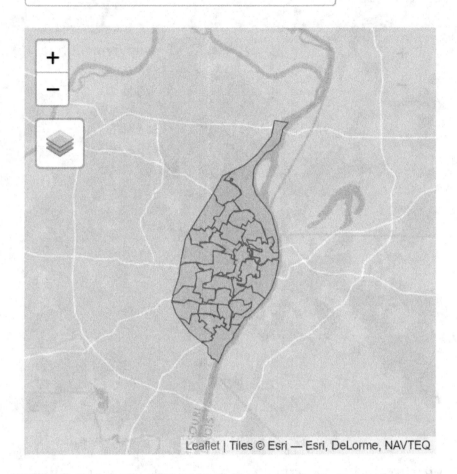

Leaflet | Tiles © Esri — Esri, DeLorme, NAVTEQ

FIGURE 10.3 St. Louis Ward Maps Shiny app showing rates of unemployment.

```
    selected = stl_plot_choices[1]
  )
)

# creating the server
server <- function(input, output) {
  # this is the plot that will load
```

St. Louis Maps

FIGURE 10.4 St. Louis Ward Maps Shiny app showing St. Louis Wards.

```
plotOutput("plot")
output$plot <- renderPlot({
  ggplot(female_low_unemployment, aes(
    x = total,
    y = reorder(female_jobs, total)
  )) +
    geom_col() +
    labs(
```

```
          title = "Top 10 female jobs in areas with low unemployment",
          x = "Total",
          y = "Occupation"
        )
  })

  observeEvent(input$plots, {
    if (input$plots == stl_plot_choices[2]) {
      output$plot <- renderPlot({
        ggplot(female_low_unemployment, aes(
          x = total, y =
            reorder(female_jobs, total)
        )) +
          geom_col() +
          labs(
            title = "Top 10 female jobs in areas with low
              unemployment",
            x = "Total",
            y = "Occupation"
          )
      })
    }

    if (input$plots == stl_plot_choices[3]) {
      output$plot <- renderPlot({
        ggplot(male_low_unemployment, aes(
          x = total,
          y = reorder(male_jobs, total)
        )) +
          geom_col() +
          labs(
            title = "Top 10 male jobs in areass with low
              unemployment",
            x = "Total",
            y = "Occupation"
          )
      })
    }
  })
}

# running the app
shinyApp(ui = ui, server = server)
```

In this app, in the UI function, the `stl_plot_choices` variable stores the drop-down menu options which are "Choose a plot," "Occupation with the lowest unemployment (female)," and "Occupation with the lowest unemployment (male)" which is called in the `selectInput` function. The option that is selected by default is "Choose a plot." On the server side, the `plotOutput` function displays the plot, which is creatively called "plot." Through the `renderPlot` function, the default plot that is loaded before any selection is made will be the "Occupation with the lowest unemployment (female)" plot. The `observeEvent` plot allows the plot that is being displayed in the app to be updated based on the user selection. To enable this, we have two `if` statements in which if the user chooses `stl_plot_choices[2]` (Occupation with the lowest unemployment (female) plot), then it will display or `stl_plot_choices[3]` Occupation with the lowest unemployment (male) plot), then the specific `renderPlot` function will run.

Now that we are finished with this app, run the app to see how it looks.

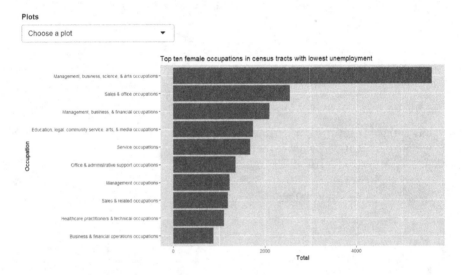

FIGURE 10.5 Default shiny plot app.

10.8   Integrating Shiny Apps into a Flexdashboard

Now we are going to put these Shiny apps into the flexdashboard that we created in the previous chapter. We will do this by putting the UI function in the sidebar and putting the server function in the second column that shows

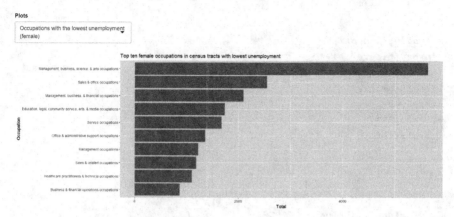

FIGURE 10.6 Plot of occupations with the lowest unemployment(female).

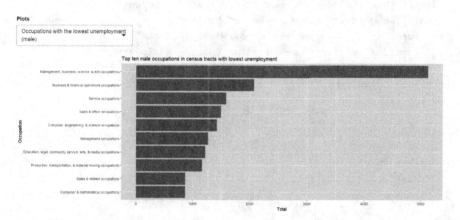

FIGURE 10.7 Plot of occupations with the lowest unemployment (male).

our output. First, we will do it with the Shiny mapping app. Let's put the UI in the sidebar of the St. Louis Ward Maps page:

```
---
title: "St. Louis Shiny Dashboard"
output: flexdashboard::flex_dashboard
runtime: shiny
---




```{r setup, include=FALSE}
knitr::opts_chunk$set(echo = TRUE)
```
```

```
# loading libraries
library(tidyverse)
library(flexdashboard)
library(tmap)
library(sf)
library(DT)
library(shiny)
```



```{r global, include = FALSE}
aldermen_info <- read_csv("data/aldermen-contact.csv")
male_low_unemployment <- read_csv("data/male-low-unemployment.csv")
female_low_unemployment <- read_csv("data/female-low-unemployment.csv")
stl_wards <- st_read("nbrhds_wards/WARDS_2010.shp")
stl_tracts <- st_read("unemployment_tract/unemployment_tract.shp") %>%
  rename("unemployment_rate" = "unmply_") %>%
  mutate(unemployment_rate = unemployment_rate * 100)
data <- stl_wards


layers <- c(
  "Select a variable",
  "Unemployment rate by ward",
  "St. Louis Wards"
)
```


St. Louis Ward Maps
===================================

Column{.sidebar}

```{r, echo = FALSE}
selectInput(
  "var",
  "St. Louis Maps",
  choices = layers,
  selected = layers[1]
)
```
```
```

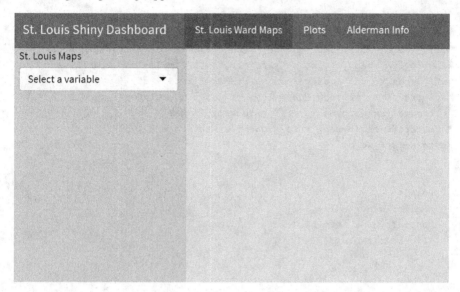

FIGURE 10.8 St. Louis dashboard with the select input bar.

Now, let's add the server function into the second column of the St. Louis Ward Maps page.

```
---
title: "St. Louis Shiny Dashboard"
output: flexdashboard::flex_dashboard
runtime: shiny
---

```{r setup, include=FALSE}
knitr::opts_chunk$set(echo = TRUE)

# loading libraries
library(tidyverse)
library(flexdashboard)
library(tmap)
library(sf)
library(DT)
library(shiny)
```


```{r global, include = FALSE}
```
```

```r
aldermen_info <- read_csv("data/aldermen-contact.csv")
male_low_unemployment <- read_csv("data/male-low-unemployment.csv")
female_low_unemployment <- read_csv("data/female-low-unemployment.csv")
stl_wards <- st_read("nbrhds_wards/WARDS_2010.shp")
stl_tracts <- st_read("unemployment_tract/unemployment_tract.shp") %>%
  rename("unemployment_rate" = "unmply_") %>%
  mutate(unemployment_rate = unemployment_rate * 100)
data <- stl_wards


layers <- c(
  "Select a variable",
  "Unemployment rate by ward",
  "St. Louis Wards"
)
```

St. Louis Ward Maps
===================================

Column{.sidebar}

```r
```{r, echo = FALSE}
selectInput(
  "var",
  "St. Louis Maps",
  choices = layers,
  selected = layers[1]
)
```

Column

```r
```{r, echo=FALSE}
# this is the map that will load
tmapOutput("map")

output$map <- renderTmap({
  tmap_mode("view")
  tm_shape(data, unit = "mi") +
    tm_polygons(zindex = 401)
})
```

```r
# we need to create a reactive variable
observeEvent(input$var, {
  if (input$var == layers[2]) {
    data <- stl_tracts
    tmapProxy("map", session, {
      tm_remove_layer(401) +
        tm_shape(data) +
        tm_fill(
          "unemployment_rate",
          title = "Umemployment rate",
          popup.vars = c("% unemployed" = "unemployment_rate"),
          id = "NAME"
        ) +
        tm_polygons(zindex = 401)
    })
  }

  if (input$var == layers[3]) {
    data <- stl_wards
    tmapProxy("map", session, {
      tm_remove_layer(401) +
        tm_shape(data) +
        tm_polygons(zindex = 401)
    })
  }
})
```

Now that we have finished integrating the Shiny app into the St. Louis Ward Maps section, we will now do the same for the Plots page. Let's first add the UI into the sidebar.

```r
---
title: "St. Louis Shiny Dashboard"
output: flexdashboard::flex_dashboard
runtime: shiny
---

```{r setup, include=FALSE}
knitr::opts_chunk$set(echo = TRUE)

# loading libraries
library(tidyverse)
```
```

FIGURE 10.9 St. Louis dashboard with the map of St. Louis Wards.

```
library(flexdashboard)
library(tmap)
library(sf)
library(DT)
library(shiny)
```


```{r global, include = FALSE}
aldermen_info <- read_csv("data/aldermen-contact.csv")
male_low_unemployment <- read_csv("data/male-low-unemployment.csv")
female_low_unemployment <- read_csv("data/female-low-unemployment.csv")
stl_wards <- st_read("nbrhds_wards/WARDS_2010.shp")
stl_tracts <- st_read("unemployment_tract/unemployment_tract.shp") %>%
  rename("unemployment_rate" = "unmply_") %>%
  mutate(unemployment_rate = unemployment_rate * 100)
data <- stl_wards


layers <- c(
  "Select a variable",
  "Unemployment rate by ward",
  "St. Louis Wards"
)
```

```
```

St. Louis Ward Maps
====================================

Column{.sidebar}

```{r, echo = FALSE}
selectInput(
  "var",
  "St. Louis Maps",
  choices = layers,
  selected = layers[1]
)
```

Column

```{r, echo=FALSE}
# this is the map that will load
tmapOutput("map")

output$map <- renderTmap({
  tmap_mode("view")
  tm_shape(data, unit = "mi") +
    tm_polygons(zindex = 401)
})


# we need to create a reactive variable
observeEvent(input$var, {
  if (input$var == layers[2]) {
    data <- stl_tracts
    tmapProxy("map", session, {
      tm_remove_layer(401) +
        tm_shape(data) +
        tm_fill("unemployment_rate",
          title = "Umemployment rate",
          popup.vars = c("% unemployed" = "unemployment_rate"),
          id = "NAME"
        ) +
        tm_polygons(zindex = 401)
    })
```

```
    }

  if (input$var == layers[3]) {
    data <- stl_wards
    tmapProxy("map", session, {
      tm_remove_layer(401) +
        tm_shape(data) +
        tm_polygons(zindex = 401)
    })
  }
})
```

Plots
===================================

Column {.sidebar}

```{r, echo = FALSE}

stl_plot_choices <- c(
  "Choose a plot",
  "Occupations with the lowest unemployment (female)",
  "Occupations with the lowest unemployment (male)"
)
selectInput(
  "plots",
  "Plots",
  choices = stl_plot_choices,
  selected = stl_plot_choices[1]
)
```

Now we will add the server to the second column which will show our main
output.

```
---
title: "St. Louis Shiny Dashboard"
output: flexdashboard::flex_dashboard
runtime: shiny
---

```{r setup, include=FALSE}
```
```

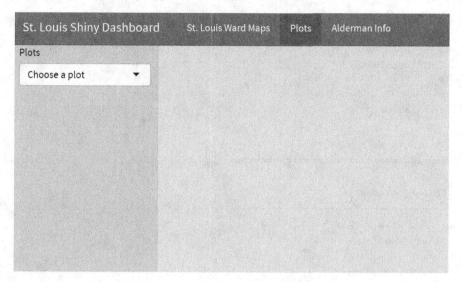

FIGURE 10.10 St. Louis Dashboard plots page with select input bar.

```
knitr::opts_chunk$set(echo = TRUE)

# loading libraries
library(tidyverse)
library(flexdashboard)
library(tmap)
library(sf)
library(DT)
library(shiny)
```

```{r global, include = FALSE}
aldermen_info <- read_csv("data/aldermen-contact.csv")
male_low_unemployment <- read_csv("data/male-low-unemployment.csv")
female_low_unemployment <- read_csv("data/female-low-unemployment.csv")
stl_wards <- st_read("nbrhds_wards/WARDS_2010.shp")
stl_tracts <- st_read("unemployment_tract/unemployment_tract.shp") %>%
  rename("unemployment_rate" = "unmply_") %>%
  mutate(unemployment_rate = unemployment_rate * 100)
data <- stl_wards


layers <- c(
```

```
   "Select a variable",
   "Unemployment rate by ward",
   "St. Louis Wards"
)
```

St. Louis Ward Maps
====================================

Column{.sidebar}

```{r, echo = FALSE}
selectInput(
  "var",
  "St. Louis Maps",
  choices = layers,
  selected = layers[1]
)
```

Column

```{r, echo=FALSE}
# this is the map that will load
tmapOutput("map")

output$map <- renderTmap({
  tmap_mode("view")
  tm_shape(data, unit = "mi") +
    tm_polygons(zindex = 401)
})


# we need to create a reactive variable
observeEvent(input$var, {
  if (input$var == layers[2]) {
    data <- stl_tracts
    tmapProxy("map", session, {
      tm_remove_layer(401) +
        tm_shape(data) +
        tm_fill("unemployment_rate",
          title = "Umemployment rate",
          popup.vars = c("% unemployed" = "unemployment_rate"),
```

```r
          id = "NAME"
      ) +
        tm_polygons(zindex = 401)
    })
  }

  if (input$var == layers[3]) {
    data <- stl_wards
    tmapProxy("map", session, {
      tm_remove_layer(401) +
        tm_shape(data) +
        tm_polygons(zindex = 401)
    })
  }
})
```

Plots
===================================

Column {.sidebar}

```{r, echo = FALSE}

stl_plot_choices <- c(
  "Choose a plot",
  "Occupations with the lowest unemployment (female)",
  "Occupations with the lowest unemployment (male)"
)
selectInput(
  "plots",
  "Plots",
  choices = stl_plot_choices,
  selected = stl_plot_choices[1]
)
```

Column

```{r, echo = FALSE}

plotOutput("plot")
output$plot <- renderPlot({
  ggplot(female_low_unemployment, aes(
```

```
      x = total,
      y = reorder(female_jobs, total)
    )) +
      geom_col() +
      labs(
        title = "Top 10 female jobss in areas with low unemployment",
        x = "Total",
        y = "Occupation"
      )
})


observeEvent(input$plots, {
  if (input$plots == stl_plot_choices[2]) {
    output$plot <- renderPlot({
      ggplot(female_low_unemployment, aes(
        x = total,
        y = reorder(female_jobs, total)
      )) +
        geom_col() +
        labs(
          title = "Top 10 female jobss in areas with low unemployment",
          x = "Total",
          y = "Occupation"
        )
    })
  }


  if (input$plots == stl_plot_choices[3]) {
    output$plot <- renderPlot({
      ggplot(male_low_unemployment, aes(
        x = total,
        y = reorder(male_jobs, total)
      )) +
        geom_col() +
        labs(
          title = "Top 10 male jobs in areas with low unemployment",
          x = "Total",
          y = "Occupation"
        )
    })
  }
})
```

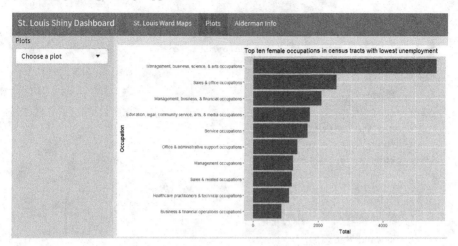

FIGURE 10.11 St. Louis dashboard showing plots.

We will not make any changes to the Alderman page.

```
---
title: "St. Louis Shiny Dashboard"
output: flexdashboard::flex_dashboard
runtime: shiny
---

```{r setup, include=FALSE}
knitr::opts_chunk$set(echo = TRUE)

# loading libraries
library(tidyverse)
library(flexdashboard)
library(tmap)
library(sf)
library(DT)
library(shiny)
```


```{r global, include = FALSE}
aldermen_info <- read_csv("data/aldermen-contact.csv")
male_low_unemployment <- read_csv("data/male-low-unemployment.csv")
female_low_unemployment <- read_csv("data/female-low-unemployment.csv")
stl_wards <- st_read("nbrhds_wards/WARDS_2010.shp")
stl_tracts <- st_read("unemployment_tract/unemployment_tract.shp") %>%
```
```

```
  rename("unemployment_rate" = "unmply_") %>%
  mutate(unemployment_rate = unemployment_rate * 100)
data <- stl_wards


layers <- c(
  "Select a variable",
  "Unemployment rate by ward",
  "St. Louis Wards"
)
```

St. Louis Ward Maps
====================================

Column{.sidebar}

```{r, echo = FALSE}
selectInput(
  "var",
  "St. Louis Maps",
  choices = layers,
  selected = layers[1]
)
```

Column

```{r, echo=FALSE}
# this is the map that will load
tmapOutput("map")

output$map <- renderTmap({
  tmap_mode("view")
  tm_shape(data, unit = "mi") +
    tm_polygons(zindex = 401)
})


# we need to create a reactive variable
observeEvent(input$var, {
  if (input$var == layers[2]) {
    data <- stl_tracts
```

```
      tmapProxy("map", session, {
        tm_remove_layer(401) +
          tm_shape(data) +
          tm_fill("unemployment_rate",
            title = "Umemployment rate",
            popup.vars = c("% unemployed" = "unemployment_rate"),
            id = "NAME"
          ) +
          tm_polygons(zindex = 401)
      })
    }

    if (input$var == layers[3]) {
      data <- stl_wards
      tmapProxy("map", session, {
        tm_remove_layer(401) +
          tm_shape(data) +
          tm_polygons(zindex = 401)
      })
    }
})
```

Plots
=====================================

Column {.sidebar}

```{r, echo = FALSE}

stl_plot_choices <- c(
  "Choose a plot",
  "Occupations with the lowest unemployment (female)",
  "Occupations with the lowest unemployment (male)"
)
selectInput(
  "plots",
  "Plots",
  choices = stl_plot_choices,
  selected = stl_plot_choices[1]
)
```


Column
```
```

```
-----------------------------------
```{r, echo = FALSE}

plotOutput("plot")
output$plot <- renderPlot({
  ggplot(female_low_unemployment, aes(
    x = total,
    y = reorder(female_jobs, total)
  )) +
    geom_col() +
    labs(
      title = "Top 10 female jobss in areas with low unemployment",
      x = "Total",
      y = "Occupation"
    )
})


observeEvent(input$plots, {
  if (input$plots == stl_plot_choices[2]) {
    output$plot <- renderPlot({
      ggplot(female_low_unemployment, aes(
        x = total,
        y = reorder(female_jobs, total)
      )) +
        geom_col() +
        labs(
          title = "Top 10 female jobs in areas with low unemployment",
          x = "Total",
          y = "Occupation"
        )
    })
  }


  if (input$plots == stl_plot_choices[3]) {
    output$plot <- renderPlot({
      ggplot(male_low_unemployment, aes(
        x = total,
        y = reorder(male_jobs, total)
      )) +
        geom_col() +
        labs(
          title = "Top 10 male jobs in areas with low unemployment",
          x = "Total",
```
```

```
          y = "Occupation"
        )
    })
  }
})
```

Alderman Info
=====================================

```{r, echo = FALSE}
datatable(aldermen_info)
```
```

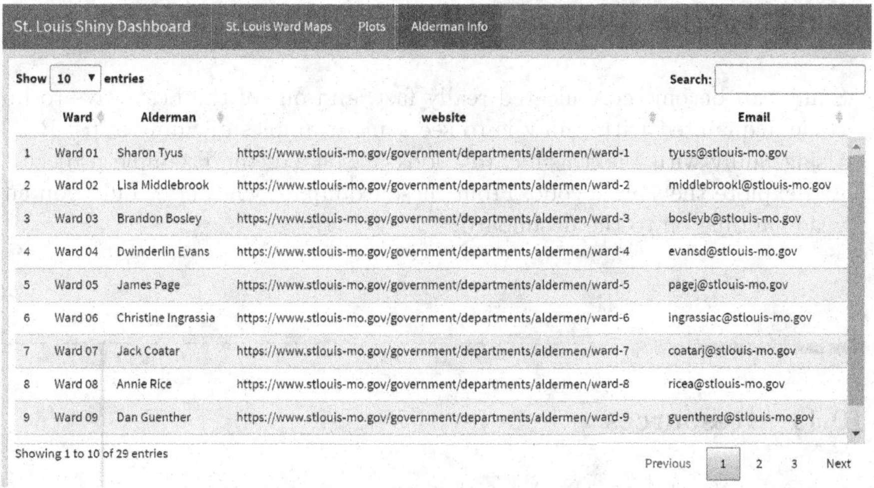


| | Ward | Alderman | website | Email |
|---|---|---|---|---|
| 1 | Ward 01 | Sharon Tyus | https://www.stlouis-mo.gov/government/departments/aldermen/ward-1 | tyuss@stlouis-mo.gov |
| 2 | Ward 02 | Lisa Middlebrook | https://www.stlouis-mo.gov/government/departments/aldermen/ward-2 | middlebrookl@stlouis-mo.gov |
| 3 | Ward 03 | Brandon Bosley | https://www.stlouis-mo.gov/government/departments/aldermen/ward-3 | bosleyb@stlouis-mo.gov |
| 4 | Ward 04 | Dwinderlin Evans | https://www.stlouis-mo.gov/government/departments/aldermen/ward-4 | evansd@stlouis-mo.gov |
| 5 | Ward 05 | James Page | https://www.stlouis-mo.gov/government/departments/aldermen/ward-5 | pagej@stlouis-mo.gov |
| 6 | Ward 06 | Christine Ingrassia | https://www.stlouis-mo.gov/government/departments/aldermen/ward-6 | ingrassiac@stlouis-mo.gov |
| 7 | Ward 07 | Jack Coatar | https://www.stlouis-mo.gov/government/departments/aldermen/ward-7 | coatarj@stlouis-mo.gov |
| 8 | Ward 08 | Annie Rice | https://www.stlouis-mo.gov/government/departments/aldermen/ward-8 | ricea@stlouis-mo.gov |
| 9 | Ward 09 | Dan Guenther | https://www.stlouis-mo.gov/government/departments/aldermen/ward-9 | guentherd@stlouis-mo.gov |

**FIGURE 10.12** St. Louis dashboard showing the aldermen information.

Congratulations, you have created a flexdasboard which integrates Shiny apps! You can refer to https://rpubs.com/sarahemlin/stl-dashboard to see the finished product.

## 10.9 Summary

The *shiny* package allows you to create interactive web applications. The foundation of Shiny applications is based on reactivity in which the output will
```

update based on the input. Two components of a Shiny app are the user interface, known as the UI and the server. The UI which displays the various elements such as selection bars, plots, maps, or tables, while the server manages the functionality of the app in terms of reactivity. We updated our flexdashboard to include various Shiny apps such as a map with drop-down menus to display either St. Louis wards or unemployment rate by ward. In addition, we added a Shiny app of a plot which also has a drop-down menu of displaying occupations in Census tracts that has the lowest unemployment by male and female occupations. We left the data table for the aldermen information unchanged.

10.10   Further Practice

- Shiny can become complicated really fast, and one of the best ways to become acquainted with Shiny is to see some examples of Shiny apps. Read Using Shiny with FlexDashboard[5], look at some of the Example projects[6], and explore the source code. From these examples, what else could you envision adding on to the dashboard?

10.11   Resources

- Shiny website: https://shiny.rstudio.com
- *Mastering Shiny*: https://mastering-shiny.org/index.html
- *tmap* package documentation: https://cran.r-project.org/web/packages/tmap/tmap.pdf

---

[5] https://rstudio.github.io/flexdashboard/articles/shiny.html
[6] https://rstudio.github.io/flexdashboard/articles/examples.html

11

Using tidymodels *to Understand Machine Learning*

11.1 Learning Objectives

1. Recall the steps in machine learning and the correlating *tidymodels* packages.
2. Describe the ways text mining is used in machine learning algorithms.
3. Describe the uses of machine learning related to employment.
4. Identify areas of potential bias in machine learning.

11.2 Terms You'll Learn

- machine learning
- algorithm

11.3 Scenario

Job seekers with whom you've worked at the library report many resumés submitted but few callbacks. You've heard many companies use resumé screening software, but you'd like to understand how that works and how it affects hiring. If there's a negative effect, you want to prepare the program participants with knowledge of how to navigate the reality of machine learning in the job application process.

11.4 Packages & Datasets Needed

```
library(tidymodels)
```

11.5 Introduction

This chapter is a brief introduction to machine learning (ML), but unlike the rest of the book, this chapter will stay reasonably high level and not include any coding. ML code requires knowledge of advanced mathematics (linear algebra); while learning to do ML is possible, the background knowledge and theory behind the steps covered in this chapter require in-depth explanations beyond the scope of this work. Learning or re-learning linear algebra takes time. However, there is still a significant benefit from understanding the process, even if you can't replicate it on your own (yet). You can still benefit from knowing the process to better interact with ML in your work and assist your library users.

The benefits of learning ML are closely related to its ubiquity in our lives. It powers many applications and systems that librarians and library users rely on for professional and personal work. Professionally, we see ML in predictive search and the search results of discovery systems and research databases we use and maintain daily. In our personal lives, our search engines queries, shopping, media platforms, and social media are all powered by ML, as are many civic (government services eligibility, job applications) and financial services (credit scores, loan approvals) we interact with regularly.

ML should merely augment human decision-making. In practice, however, we tend towards a blind faith in computers that elevates their output above our own intelligence and convinces us they can solve every problem. As information professionals, we need to understand how ML works so that we know its flaws and its promise and can communicate that to our colleagues and users. This chapter will cover what ML is and how it works, focusing on one ML R package.

11.6   What ML is

ML is a type of artificial intelligence (AI) that uses linear regression to find patterns within data, whether that data contains numbers or text. ML can surface patterns that aren't discernible to the human eye. There are two types of ML: supervised and unsupervised. For simplicity, this chapter will explain the steps of supervised ML, which means that there is a human rather than an algorithmic review of settings and decisions within the ML algorithm. An **algorithm** is a series of repeated programmatic steps, which you could think of as a formula or recipe applied to data to "predict something: how likely it is that a squiggle on a page is the letter A; how likely it is that a given customer will pay back the mortgage money a bank loans to him; which is the best next move to make in a game of tic-tac-toe, checkers, or chess" (Broussard, 2018). The steps of the algorithm repeat in the same order with any new data. Overall, ML algorithms take raw data, tidy it and then manipulate it to get the best predictive results.

ML is related to but ultimately very different in practice than statistics. With ML, the goal is to make accurate predictions using any method. As a result, ML will use any statistical model that produces more accurate predictions. While statistics use hypotheses, ML does not. In statistics, you would compare your model against random chance to see if it performed better, but in ML, the only question is which model fits the data the best.

11.7   How ML Works in R (*tidymodels*)

In the R package ecosystem, *tidymodels*[1] is the Tidyverse equivalent for ML. Like the Tidyverse, Tidymodels combines several packages that each focus on a step in the ML process. We'll cover five packages that map to the five main stages of ML.

11.7.1   Choose a Model (*parsnip*)

As previously mentioned, ML only wants the best model that fits the data, so that is the criteria by which data scientists choose their model. They need to

---

[1]https://www.tidymodels.org/

figure out what linear regression method would give the best results for the provided data. The model functions as the engine of the algorithm. Examples of models include K nearest neighbors (KNN) and root mean squared error (RMSE). The Tidymodels package to use is called *parsnip*.

11.7.2 Create a Recipe (*recipes*)

The Tidymodels *recipes* package creates the algorithm, or formula, for ML that handles data the same way every time. The recipe documents which variables the algorithm will use for predicting and the steps involved in preparing the data. Consistent data preprocessing has a significant impact on prediction accuracy. The recipe is vital for consistency because the data workflow is complex.

11.7.3 Sample the Data (*rsample*)

One crucial part of the process is dividing your dataset into training and testing datasets. The testing dataset is set aside until the end of the entire ML process to test the algorithm's accuracy. The testing dataset is used immediately with the model and the recipe. There are many decisions to make related to sample size, which all impact prediction accuracy. Too-small testing sets result in unreliable predictions because you wouldn't have tested the model fit enough. On the other hand, if the training dataset is too small, then the model fit will be poor.

11.7.4 Tune the Model (*tune*)

Because ML aims to get the best possible model that fits the data at hand, a sizable portion of the process is devoted to adjusting or tuning the model to get a better fit. Tidymodels uses the *tune* package for model tuning. Changing the algorithm parameters and hyperparameters in different ways can improve model fit. Resampling is one example of this, which reconfigures the training dataset in various ways to determine which produces the best results.

11.7.5 Measure the Fit of the Model (*yardstick*)

The last package for this chapter is *yardstick*, which measures mathematically how accurately the chosen model fits the data. This package would be run on the testing dataset (reserved from the beginning and not used to develop the algorithm) as a final validation step to see how well it predicts. "Good" models might fit 80% of their data points. Depending on the algorithm's use in decision-making, 80% accuracy might be benign or harmful. For example, 75% accuracy on which movie appears in the "Suggested" tab on your streaming platform that you definitely want to watch has a different impact than an algorithm used to predict whose children should be removed by the state that is wrong about 25% of the time (Eubanks, 2018).

11.8 Problems with Machine Learning

Implicit in the above ML steps is that all ML algorithms are created by humans using data created by humans. Keeping this top of mind helps us remember that AI has no magic. Even if a commercial library database uses an ML algorithm that we can't see (called a "black box" algorithm), we know that their algorithm uses the same preceding steps. Knowing the ML process means we can understand where potential bias or error is likely to sneak in, allowing us to critique and analyze an ML algorithm.

The impact of model accuracy is just one area where problems can arise in the real-world application of ML algorithms. Human bias and problems with the data can combine and compound in ways that can be invisible to data scientists, producing unintended negative consequences for ML-augmented decision-making.

11.8.1 The data

The root cause of many ML problems is often the data it ingests. Sometimes, the data itself is too small or too problematic. Quite often, the data we need to solve a problem doesn't exist, so we use the data we can find to act as a proxy for the information we don't have. This generalization is understandable, but again, the validity and credibility of an ML-based decision deserve careful thought to avoid unintended negative consequences.

Contrasting online shopping data with social services data illustrates this point well. Companies want us to buy early and often and use data to predict what we will buy. Online merchants track every click and keystroke on their shopping platforms, collecting tremendous data to drive product suggestions and email marketing campaigns. ML algorithms can't infer intent, though, so you might purchase a gift and then get a related purchase suggestion that's irrelevant. The downstream consequences of a wrong prediction might result in minor annoyance. Data collected by child welfare agencies is entirely different. These agencies aim to predict which children among the entire population are at risk for neglect or abuse (Eubanks, 2018).

Data is collected by and created by humans, encoding all of our biases and societal problems: "if the data comes from humans, it will likely have a bias in it"(Shane, 2019). An excellent example of this problem is zip codes. Zip codes (postal codes) are used as a clean data point to say where someone lives or where an event occurs. However, in the United States, zip codes are an unintended proxy for race (O'Neil, 2016). For decades a process called redlining forced non-white populations into particular geographic areas, often using natural barriers like highways. At the same time, white people could live and buy houses anywhere they wanted (Rothstein, 2017). Real estate isn't a mobile possession, leaving areas with low access to services and jobs synonymous with their non-white populations. Even though redlining is illegal, racial disparities between zip codes persist.

11.8.2 Assumptions

The assumptions we make about the data we have is another potential problem. Essentially, an ML algorithm assumes that the data at hand can answer the question posed by the researcher. Not all of these assumptions are harmful; even if benign, practitioners need to be aware of the underlying assumptions in the data, the algorithm, and the analysis. Using one set of data as a proxy for data you don't have or doesn't exist should be a red flag that assumptions are present(O'Neil, 2016). A common belief is when we think an algorithm can replace tasks usually performed by humans. A good proving ground for this is library discovery systems, which use text as its data input to predict which library resources are most relevant for each search query. Librarians are experts in answering reference questions, but library discovery systems aren't a 1:1 replacement. We must remember that "search engines are best when retrieving information about the mundane business of everyday life...but library discovery systems were explicitly designed to deal with topics from the ordinary to the complex" (Reidsma, 2018). There's an assumption baked into the acquisition of these systems that they will be objective and accurate when they are often wrong and always opaque about how relevancy is determined. People created ML algorithms that encode their bias within the math.

11.8.3 Sampling

Dividing your dataset into different samples for training and testing your algorithm is vital in ML, but the division itself can be problematic. For example, if your dataset is too small, your training set might produce a model that doesn't work on the testing dataset. Alternatively, irregularly distributed data results in inaccurate models and predictions even if the dataset is large enough. An example is earthquakes, which are not evenly distributed and are very hard to predict (Silver, 2012).

11.8.4 Tuning & Measuring Fit

The final step in ML is tuning the model and measuring the fit, which says how accurate its predictions are. The greatest danger in tuning is overfitting the model. These tradeoffs speak to another potential ML problem: if the model fits the data perfectly, it might predict horribly in real life. For example, using data on one town's home sale prices to predict another town's sale prices may or may not work. Another example with more at stake is overfitted models, such as a self-driving car trained on images with grass on both sides of the road that can't process arriving at the start of a bridge (Shane, 2019).

11.9 Machine Learning in Employment

The reality of job searches today is that when "you submit a job application or resumé via an online job site, an algorithm generally determines whether you meet the criteria to be evaluated by a human or whether you're rejected outright" (Broussard, 2018). All job seekers in your outreach program must deal with this, so it's essential to understand what is happening and why.

Resumé screening algorithms utilize text mining and apply ML to words rather than numbers. The goal is to parse the words in a resumé or application to predict which candidates will be most successful; those predicted to do well get passed on to the next round in the hiring process. When we think of the words included in an employment application, names, industry keywords, company names, and educational institutions are listed. All of the previously enumerated problems with ML are present in a screening algorithm, and there is no recourse for an individual denied an interview for what they think is an illegitimate reason.

These algorithms use words that appear in "successful" applications with current applications. Yet if a company has hired many people who are similar in name, education, or background, then the algorithm will predict that only people like them will be successful. Because we believe computers are objective, algorithms are often undisputed, and our decisions encode the disparities and inequities in our current environment. The lexicons used in textual ML are of particular concern because models trained on lexicons that originated from crawling the internet can have word associations that reflect the worst parts of our culture back to us. Hazardous word associations can manifest in automated restaurant reviews that consistently demote Mexican restaurants because the training data contained a strong word association between "Mexican" and "illegal," a term with a negative connotation that carries over to what the algorithm associates as a related word (Shane, 2019). While we can't say if a particular algorithm is making negative associations where it shouldn't be, we know that using lexicons based on the internet is popular because the internet holds a vast number of words, and crawling it to create a dataset is free.

11.10   Summary

ML is ubiquitous, but we need to understand how it works because it's so intertwined with our personal and professional lives. Supervised ML uses data to fit a model that predicts results, where people decide to set sample sizes, set model parameters, and adjust the model's fit. Because people make the data and they make the algorithms, they are both fallible, and placing blind faith in algorithmic predictions as "truth" is unwarranted. There are many places where bias can creep into an ML algorithm, making it critical that we probe the origins of the analyzed data and unpack assumptions therein.

11.11   Further Practice

- Work through the examples in each chapter of *Supervised machine learning for text analysis in R* by Emil Hvitfeldt & Julia Silge: https://smltar.com/

11.12   Additional Resources

- Tidymodels website: https://tidymodels.org
- *You look like a thing and I love you* by Janelle Shane
- *Masked by trust* by Matthew Reidsma
- *Automating inequality* by Virginia Eubanks
- *Weapons of math destruction* by Cathy O'Neil
- *Artificial unintelligence* by Meredith Broussard

12

Conclusion

12.1 Wrapping Up...for Now

Congratulations on reaching the end of this book! The book's purpose was to give you a basic foundation in R, which you can use for library-related work. Your work will benefit the fictional you and your fictional partners in our storyline in terms of gaining more knowledge and experience with using R and applying it for a particular library-based purpose. This knowledge and skills will also benefit you in the real world. To recap what you have done:

1. Become familiar with the RStudio IDE
2. Learn how to do basic data cleaning and data scrubbing tasks with the *dplyr* package.
3. Make a basic plot with the *ggplot2* package.
4. Do web scraping with the *rvest* package.
5. Create a static and interactive map with the *tmap* package.
6. Do text mining operations such as tokenization, sentiment analysis, and tf-idf function.
7. Create a basic R markdown document.
8. Create a flexdashboard using R Markdown.
9. Create Shiny apps within the flexdashboard.
10. Have a conceptual understanding of machine learning using the *tidymodels* package.

The final deliverables you created through the scenario are an R Markdown document and flexdashboard that integrates a Shiny app in order to summarize the wards in St. Louis with the lowest and highest unemployment along with the top ten occupations in each of these wards.

DOI: 10.1201/9781003218012-12

12.2   Where Do You Go from Here?

You're probably wondering what's next. This book focused on the breadth of
R, not the depth. Where you go from here depends on what you want to focus
on moving forward. We recommend the seminal text *R for Data Science*[1] by
Garrett Grolemund and Hadley Wickham, for a more robust foundation with
R. *R for Data Science*, more commonly referred to as R4DS, will give a more
in-depth overview of such topics as *dplyr*, *ggplot2*, and tidy data principles.

Suppose you want to learn more about *tmap* and to handle spatial data in R in
general. In that case, we recommend *Geocomputation in R*[2] by Robin Lovelace,
Jakub Nowosad, and Jannes Muenchow. This book gives a comprehensive
overview of the types of spatial data, operations related to the different types
and components of spatial data, making maps with R with various packages
other than *tmap*, and applications in various fields.

When it comes to web scraping and text mining, you can't go wrong with *Web
Scraping with R*[3] by Steve Pittard and *Text Mining with R*[4] by Julia Silge and
David Robinson. *Web Scraping with R* provides several real-world examples for
web scraping using APIs and guidance on dealing with websites that return
XML and JSON when attempting to scrape them. With *Text Mining with
R*, there is a more in-depth discussion on topics such as tf-idf and sentiment
analysis and covers subjects not included in this text, such as n-grams and
correlations.

An excellent text to refer to if you want to learn more about *rmarkdown* and
flexdashboard is *R Markdown: The Definitive Guide*[5] by Yuhui Xie, J.J. Allaire,
and Garrett Grolemund. It gives more detail about R Markdown syntax and
FlexDashboard and real-world examples and other outputs you can create
with R Markdown, such as presentations, document templates, and websites.

We only covered the surface of *shiny* in this book. For a more comprehensive
discussion on Shiny, *Mastering Shiny*[6] by Hadley Wickham is a good resource,
which gives a more in-depth discussion of reactivity and best practices for
making Shiny apps. Another topic that we didn't go in-depth with is how to
do machine learning with the *tidymodels* package. *Tidy Modeling with R*[7] by

---

[1] https://r4ds.had.co.nz/
[2] https://geocompr.robinlovelace.net/
[3] https://steviep42.github.io/webscraping/book/
[4] https://www.tidytextmining.com/index.html
[5] https://bookdown.org/yihui/rmarkdown/
[6] https://mastering-shiny.org/
[7] https://www.tmwr.org/

Max Kuhn and Julia Silge will expose you to both the basics and the basics of tidy modeling through case studies using housing data.

12.3 Data Management: Never Do Work Without It

We provided many resources to further your knowledge on the various topics covered in this book. However, it is also crucial that you implement sound data management practices that allow your data to be FAIR which, to reiterate, means that your data is findable, accessible, interoperable, and reusable (see Chapter 1 for further reference). One example of sound data management principles includes having a README file that mentions elements such as the name of your project, contact information, file, and variable listings, along with an explanation of the variables. Another example is ensuring that your variable names do not have special characters and that an underscore or hyphen is used instead of spaces since spaces can cause problems when reading in a file or variables. A good resource for data management is *Data Management for Researchers*[8] by Kristin Briney. By implementing sound data management practices, you're helping others understand what you are doing and the future you as well when you have to revisit a project that you haven't worked on in a while.

12.4 Final Send-off on Your Data Science Journey

This book has launched your data science journey, but it is just the beginning. We hope that through this book, you learned basic data science concepts such as data scrubbing and data visualization, and that you gained confidence by working through the scenarios for each chapter. We hope you can see how some of the things you learned can be applied to your daily work and find an interesting topic you would explore more in-depth. Good luck on your data science journey, and while there might be turns and road bumps on the way, keep on going and don't stop learning!

---

[8]https://pelagicpublishing.com/products/data-management-for-researchers-briney#

A

Dependencies

A.1 iOS Dependencies

Homebrew[1] is an application needed by iOS operating systems to install additional software/packages. Follow the instructions on their website to install Homebrew on your Mac. The password requested is the one used to log in to your computer.

- go to https://brew.sh and copy their install code for Homebrew for Mac
- open the terminal and paste that code; follow any prompts
- if it was installed a while ago, then `brew update` is needed
- once installed, run the following commands (one at a time):
  - `brew install udunits`
  - `brew install gdal`
  - `brew install proj` **this might already be installed

For Mac users on Big Sur who encounter problems with the geospatial package installations, try: install.packages(c("rgdal","sf"), "https://mac.R-project.org") `"install.packages(c("rgdal","sf"), "https://mac.R-project. org")"` and answer yes to first prompt, but no to compilation from source

A.2 Windows Dependencies

In order to install and compile packages, Windows users need to install *Rtools42*[2]. Once *Rtools42* is installed, there's no need to install the *Rtools* package.

---

[1] https://brew.sh/
[2] https://cran.r-project.org/bin/windows/Rtools/rtols42/rtools.html

A.3   Package Dependencies for This Book

Some of the packages used in this book have other packages that they depend on. You'll usually find this out in one of two ways. Either the dependencies will be installed with the package you're installing and you'll see messages notifying you in the console, or else the `install.packages()` function will return an error and let you know which package(s) is missing. You'll then want to install that package(s) first before returning to install the initial package you were trying to install.

B

Additional Skills

B.1 Using the Shell or Command Line on Your Computer

If you don't know how or don't feel comfortable using the command line on your computer, we recommend attending a session of the Carpentries' UNIX Shell class, or working through the course materials online[1].

B.2 Using GitHub & Git

While version control systems like Git are widely used in academia, all librarians can benefit from using them to back up their code and collaborate with others. If Git and GitHub are new to you, we recommend the Carpentries' Version Control with Git class. Check their website for upcoming classes near you, or work through the course materials online[2].

B.3 Troubleshooting Package Installation Problems

Sometimes, depending on your particular computer OS and package installations, installing the geospatial packages will not work. We've found that using: `install.packages("https://cran.r-project.org/bin/macosx/contrib/4.2/stars_0.5-5.tgz", repos = NULL, type = .Platform$pkgType)`, and replacing the URL in that code with a link to the appropriate binary version of

---

[1]https://librarycarpentry.org/lc-shell/
[2]https://swcarpentry.github.io/git-novice/

the package you need to download from CRAN will overcome any difficulties with the spatial packages. Repeat with new package links until all packages have been installed.

There are lots of times when the code you have will return an error message. When that happens, most users copy the code and paste it into a search box and see what the internet returns. Sites like StackOverflow[3] and the Posit Community forum[4] are common places to ask questions about problems you're having or read answers to other people's questions that might be similar to yours. You might need quite a bit of trial and error before you're able to fix your problem(s); take comfort in the thousands and thousands of posts on those sites which show that other people have problems with their code, too.

---

[3] https://stackoverflow.com/
[4] https://community.rstudio.com

Bibliography

Broussard, M. (2018). *Artificial unintelligence.* MIT Press, Cambridge, Massachusetts. ISBN 978-0262537018.

Eubanks, V. (2018). *Automating equality.* Picador, New York. ISBN 978-1250215789.

Hu, M. and Liu, B. (2004). Mining and summarizing customer reviews. In *Proceedings of the Tenth ACM SIGKDD International Conference on Knowledge Discovery and Data Mining*, KDD '04, page 168–177. ACM, New York, NY, USA.

Matalon, Y., Magdaci, O., Almozlino, A., and et al. (2021). Using sentiment analysis to predict opinion inversion in tweets of political communication. *Scientific Reports*, (11):7250.

Mohammad, S.M. and Turney, P.D. (2013). Crowdsourcing a word-emotion association lexicon. *Computational Intelligence*, 29(3):436–465.

Navarro,, H.W.D. and Pedersen, T.L. (2020). *ggplot2: Elegant graphics for data analysis.* Springer, New York.

Network, N.Y.T.D. (2022). Get started. https://developer.nytimes.com/get-started/.

O'Neil, C. (2016). *Weapons of math destruction.* Broadway Books, New York. ISBN 978-0553418835.

Reidsma, M. (2018). *Masked by trust.* Litwin, Sacramento, California. ISBN 978-1634000833.

Robin Lovelace, and Jakub Nowosad, J.M. (2019). *Geocomputation in R.* CRC Press, Boca Raton, Florida. ISBN 9781351396905.

Rothstein, R. (2017). *Color of law.* Liveright, New York. ISBN 978-1631494536.

Shane, J. (2019). *You look like a thing and I love you.* Little, Brown, and Company, New York. ISBN 978-0316525244.

Silver, N. (2012). *Signal and the noise*. Penguin Books, New York. ISBN 978-0143125082.

Wickham, H. (2021). *Mastering Shiny*. O'Reilly Media, Sebastopol, California. ISBN 978-1492047384.

Wickham, H. and Grolemund, G. (2016). *R for data science*. O'Reilly Media, Sebastopol, California. ISBN 978-1491910399.

Wilkinson, L. (2005). *The grammar of graphics*. Springer, New York. ISBN 978-0-387-28695-2.

Index

Printed in the United States
by Baker & Taylor Publisher Services